RAISING PUBLIC AWARENESS OF ENGINEERING

Lance A. Davis and Robin D. Gibbin, Editors

NATIONAL ACADEMY OF ENGINEERING
OF THE NATIONAL ACADEMIES

The National Academies Press
Washington, D.C.
www.nap.edu

THE NATIONAL ACADEMIES PRESS • **500 Fifth Street, N.W.** • **Washington, DC 20001**

NOTICE: This publication has been reviewed according to procedures approved by the National Academy of Engineering report review process.

Funding for the activity that led to this publication was provided by the Elizabeth and Stephen Bechtel, Jr., Foundation. Any opinions, findings, conclusions, or recommendations expressed in this publication are those of the author(s) and do not necessarily reflect the views of the organizations or agencies that provided support for the project.

International Standard Book Number 0-309-08624-8

Additional copies of this report are available from the National Academies Press, 500 Fifth Street, N.W., Lockbox 285, Washington, DC 20055; (800) 624-6242 or (202) 334-3313 (in the Washington metropolitan area); Internet, http://www.nap.edu

Copyright 2002 by the National Academy of Sciences. All rights reserved.

Printed in the United States of America

THE NATIONAL ACADEMIES
Advisers to the Nation on Science, Engineering, and Medicine

The **National Academy of Sciences** is a private, nonprofit, self-perpetuating society of distinguished scholars engaged in scientific and engineering research, dedicated to the furtherance of science and technology and to their use for the general welfare. Upon the authority of the charter granted to it by the Congress in 1863, the Academy has a mandate that requires it to advise the federal government on scientific and technical matters. Dr. Bruce M. Alberts is president of the National Academy of Sciences.

The **National Academy of Engineering** was established in 1964, under the charter of the National Academy of Sciences, as a parallel organization of outstanding engineers. It is autonomous in its administration and in the selection of its members, sharing with the National Academy of Sciences the responsibility for advising the federal government. The National Academy of Engineering also sponsors engineering programs aimed at meeting national needs, encourages education and research, and recognizes the superior achievements of engineers. Dr. Wm. A. Wulf is president of the National Academy of Engineering.

The **Institute of Medicine** was established in 1970 by the National Academy of Sciences to secure the services of eminent members of appropriate professions in the examination of policy matters pertaining to the health of the public. The Institute acts under the responsibility given to the National Academy of Sciences by its congressional charter to be an adviser to the federal government and, upon its own initiative, to identify issues of medical care, research, and education. Dr. Harvey V. Fineberg is president of the Institute of Medicine.

The **National Research Council** was organized by the National Academy of Sciences in 1916 to associate the broad community of science and technology with the Academy's purposes of furthering knowledge and advising the federal government. Functioning in accordance with general policies determined by the Academy, the Council has become the principal operating agency of both the National Academy of Sciences and the National Academy of Engineering in providing services to the government, the public, and the scientific and engineering communities. The Council is administered jointly by both Academies and the Institute of Medicine. Dr. Bruce M. Alberts and Dr. Wm. A. Wulf are chair and vice chair, respectively, of the National Research Council.

www.national-academies.org

Preface

In April 2001, The National Academy of Engineering (NAE), with funding from the Elizabeth and Stephen Bechtel, Jr., Foundation, initiated a new project as part of its Public Understanding of Engineering Program. The project was informed by public opinion surveys that indicated that the American public has minimal understanding of engineering and engineers and little appreciation for what they do. A survey/questionnaire was developed in conjunction with outside consultants, Market Research Bureau and McMahon Communications, to create an inventory of current outreach programs for improving public awareness of engineering and to better understand their effectiveness or lack of effectiveness. The survey was administered and the results were collected by the consultants. To review the results of the survey/questionnaire, the NAE formed the Committee on Public Awareness of Engineering (CPAE), composed of a group of distinguished citizens interested in and/or involved in the engineering community. Based on their review, the committee made recommendations for the engineering community's future approach to public awareness activities. This report is a report from the NAE to the engineering community based on a synthesis of the consultants' report of the survey, background information developed by the NAE, and the recommendations of the CPAE.

Wm. A. Wulf
President
National Academy of Engineering

Acknowledgments

This report has been reviewed in draft form by individuals chosen for their diverse perspectives and technical expertise, in accordance with procedures approved by the National Academy of Engineering report review process.

The purpose of this independent review is to provide candid and critical comments that will assist the institution in making its published report as sound as possible and to ensure that the report meets institutional standards for objectivity, evidence, and responsiveness to the study charge. The review comments and draft manuscript remain confidential to protect the integrity of the deliberative process. We wish to thank the following individuals for their participation in the review of this report: Albert A. Dorman, AECOM; Samuel C. Florman, Kreisler Borg Florman Construction Company; Henry Kressel, E.M. Warburg, Pincus & Company, LLC; Susan Staffin Metz, Stevens Institute of Technology; Robert L. O'Rourke, California Institute of Technology; and Winfred M. Phillips, University of Florida.

Although the reviewers listed above have provided many constructive comments and suggestions, they were not asked to endorse the conclusions or recommendations nor did they see the final draft report before its release. The review of this report was overseen by Harold Forsen, National Academy of Engineering. Appointed by the National Academy of Engineering Executive Office, he was responsible for making certain that an independent examination of this report was carried out in accordance with institutional procedures and that all review comments were carefully considered. Responsibility for the final content of this report rests entirely with the institution.

Contents

Executive Summary — 1

1 Introduction — 6
   Importance of Engineering, 6

2 Public Awareness Today — 10
   Follow-on Survey, 10
   Conclusion, 17

3 The National Academy of Engineering Survey of Public Awareness of Engineering Outreach Programs — 18
   Ojectives and Methodology, 19
   The Current Situation, 22
   Interpretation of Findings, 38
   Messages, 43

4 Recommendations — 47
   Setting the Stage, 47
   Goal: Improved Public Awareness of Engineering, 48
   Next Steps, 51

Appendixes

A	Engineering Enrollments	55
B	Sample Mission Statements	62
C	Selected Outreach Programs	64
D	Committee on Public Awareness of Engineering	71
E	Engineering Communications, Education, and Outreach Questionnaire	74
F	List of Organizations Responding to NAE Inventory Questionnaire	93

Executive Summary

In the twentieth century, engineers and engineering made disproportionate contributions, in comparison to their numbers and the credit received, to the design and development of the infrastructures and technologies that support the nation's global competitiveness, security, and standard of living. And they will continue to do so in the twenty-first century. Yet as our lives become more and more dependent on technological marvels, we and our elected representatives understand less and less about it. Most American citizens are poorly equipped to engage in public debate about technology-related issues that may affect their lives; our elected representatives are also poorly equipped to make decisions about technology-based policy issues. To compound the problem, the K-12 educational system does a poor job of teaching math and science to children (and rarely teaches engineering and technology at all). Thus, new generations of engineers cannot be taken for granted. In fact, engineering graduations have been flat for the last 10 years.

To address these issues, many organizations in the engineering community have undertaken many programs to try to improve the public awareness and public understanding of engineering. Unfortunately, public opinion surveys indicate that these programs have had little or no measurable impact on public perceptions of engineering.

A significant improvement in public awareness will require coordinated efforts by engineering organizations presenting consistent messages about the nature and value of engineering. To help the engineering community improve its individual and collective efforts, the National Academy of Engineering commissioned a survey/questionnaire to identify the range of current public awareness activities, and to collect information on their goals, target audiences, messages, finances, and so on. This report summarizes the results of that survey and offers recommendations from the Committee on the Public Awareness of Engineering, a committee of distinguished members of the engineering community.

SUMMARY OF THE SURVEY OF THE ENGINEERING COMMUNITY

A great many outreach, communications, and educational activities are being conducted by engineering organizations. Many of these organizations sponsor more than one program and have engaged in these activities for many years. The total expenditures for all organizations that provided figures are about $265 million. We can estimate total expenditures for all organizations that reported they had outreach programs by extrapolating the mean budget for those which provided funding figures. Thus, total expenditures could be as high as $400 million, not including the value of volunteers' time. Organizations involved in these activities are committed to continuing the programs for the foreseeable future, and many plan to initiate new programs as well.

Although these activities are diverse, many of them are intended to inspire young people (K-12) to pursue careers in engineering by introducing them to engineering and by stressing the importance of math and science. These programs convey a simple message that math and science are fun and can lead to rewarding, challenging, fun, exciting, creative careers in engineering.

All types of organizations surveyed, colleges and universities, engineering societies, museums, and national laboratories, have outreach programs. Most current activities are local or regional in scope. Only a few are national programs, such as National Engineers Week, although it is typically conducted as a series of local projects. Most activities involve working with local schools, mentoring programs, and similar activities. In addition, contact between engineers and students is intermittent and temporary, sometimes just once a year. Although these "grassroots" activities appear to be well received, they have not had a measurable impact on public awareness of engineering on a national scale, as measured by public attitude surveys (Chapter 2).

Most of these programs deliver messages on similar themes. There are recruitment messages—engineering is a fun, creative, exciting, important career; "math and science are fun" messages; and "engineers are important and contribute to the quality of life, economy, environment" messages. Although the wording differs from program to program, these themes appear over and over.

Organizations that sponsor outreach activities are very proud of their efforts and believe their programs are successful. However, their measurements of success tend to focus on short-term processes and tactics rather than long-term outcomes. The organizations who believed their programs had not been successful attributed this to a lack of resources. In fact, no single program can be cited, based on objective measures, as being particularly effective. National Engineers Week is the most highly visible and entrenched program, and many respondents considered it is a good program; but it does not have universal reach. Other programs cited for effectiveness were sponsored by the National Society of Professional Engineers, the American Society of Civil Engineers, the National Academy of Engineering, and the American Association of Engineering Societies.

When asked if they would support coordinated efforts, most, but not all, answered yes. Many respondents noted that bringing together all of the specialized societies and trade associations (both of which tend to be very territorial) in one coordinated campaign would be difficult. However, most respondents believed coordinated efforts would be more efficient and more effective because they would deliver consistent messages, and most of the organizations were prepared to participate, or at least consider participating. The primary reasons they gave for supporting coordinated efforts can be summed up as improving the image of engineers and engineering among the public. A few felt that coordinated efforts would complement, rather than replace their programs, and some felt that their own resources (money and staffing) were already stretched too thin. Respondents who did not believe a coordinated effort is necessary thought it would be difficult to run and difficult to agree on messages or objectives. When asked what the messages of a large-scale campaign should be, most suggested that they should emphasize the importance of engineers to society and should promote engineering as a career.

RECOMMENDATIONS

A coordinated campaign to improve public understanding of engineering will require both short-term and long-term actions. The short-term focus should be on maintaining and increasing the public awareness of engineering through public relations and public affairs (PR/PA) activities. Long-term activities should focus on changes in the educational curriculum and improved teaching of math and science in elementary and secondary school. The engineering community is already engaged in PR/PA activities and educational interventions, but it needs to be better coordinated to ensure critical mass and better measured to ensure effectiveness.

If students are successfully engaged by math, science, engineering, and technology in grammar school and their interest can be sustained through secondary school, the goals of having a more technologically literate populace and students educationally equipped to choose an engineering track in college will be in hand. However, even assuming present education activities can be leveraged successfully on a national scale, it will take a generation to see the long-term outcomes. Views of engineering held by the public are based on decades of information, misinformation, or lack of information, and will be difficult to change. Young people, beginning school, obviously, do not have these perceptions. They are clean slates, and their perceptions can be molded more easily. However, the engineering community and the nation can not afford to wait for this process to unfold, nor presume its ultimate success, so more effective PR/PA efforts now and in the future are essential.

- *The engineering community should develop nationally coordinated public relations and public affairs grassroots efforts to improve the public awareness of engineering.* Present efforts are largely local in effort and have not demonstrated impact on a national scale.
- *The engineering community should agree on consistent messages (e.g., slogans, catchwords, images, etc.) for those campaigns that have been developed through rigorous testing to ensure their effectiveness. These messages should then be used throughout the community.* The information provided in the survey/questionnaire includes some promising messages, or a basis for developing effective messages.
- *Metrics and evaluation criteria for future programs must be established.* Decades of well-intentioned and enthusiastic outreach at the grassroots level have had little impact on engineering enrollments or public attitudes. Evaluation is important for two reasons: so that programs can demonstrate results, which in turn will safeguard their credibility and help garner ongoing support from the engineering community; and so that ineffective programs can be modified for greater impact.
- *A mechanism should be established, e.g., a web site, to share public awareness of engineering activities with the entire engineering community.* Appendix C lists selected outreach programs that appear to demonstrate some evaluation and impact. These outreach programs should be shared with the community to allow others to use and build on them. One of the goals of the project was to identify best practices. However, since most organizations do not evaluate their programs with objective criteria, it was not possible to select "best in class" programs.
- *Outreach activities that organizations are currently engaged in should be continued and encouraged.* Renewed efforts should be made to develop objective outcome measures to ensure their effectiveness, but, at a minimum, present programs lend credibility to the commitment of the profession to improving technology literacy of the public and the education system. Some of them may be leveraged as part of nationwide efforts.
- *The media should be educated about engineering issues and the engineering community should place resources at their disposal.* The media are very influential, and there is much room for educating them on the right way to talk about engineering and technology issues. The profession should also ensure that it can support inquiries from the media, providing them with spokespeople and experts.
- *New programs for children should also be developed to show how engineering is integrated into all aspects of society.* Saturday morning

television, movies, other popular media, and museums should be strongly pursued to incorporate engineering, math and science messages.

Several of the above recommendations apply to educational intervention as well as to public relations and public affairs efforts. For the specific case of the education system:

- *The engineering community should create a blue-ribbon council of representatives of the engineering, education, and public policy communities to develop an action plan for improving math, science, engineering, and technology education.* Math and science curricula must be re-assessed and new curricula designed to encourage long-term study and to introduce engineering and problem solving in the early grades. Initial efforts should be focused on younger children who are less likely to have negative perceptions toward math and science. The state of Massachusetts, through the efforts of Tufts University, the Museum of Science, and other organizations and policy makers, has already embarked on a concerted effort that involves setting standards, redesigning curricula, and retraining teachers. Such approaches could eventually solve the pipeline issue and also make great improvements in increasing technical literacy and changing the image of engineers and engineering.
- *Curricula redesign efforts should be supported through efforts targeting opinion leaders and public policy leaders.* Engineering's voice should be heard in public policy, through direct communications to opinion leaders and policy makers. Engineering societies and many companies already have lobbying efforts to ensure that their interests are represented in policy discussions. Similar efforts should be made to ensure that policy makers are educated about the importance of changing curricula.

1

Introduction

Engineering is the profession in which a knowledge of the mathematical and natural sciences gained by study, experience, and practice is applied with judgment to develop ways to utilize, economically, the materials and forces of nature for the benefit of mankind.

Accreditation Board for Engineering and Technology, (ABET)

IMPORTANCE OF ENGINEERING

Who cares about public awareness of engineering? This question can be answered rather simply—The engineering community cares. They care because they are chagrined to recognize that the public does not understand the contributions of engineers and engineering to their quality of life; that the public is not equipped to engage in public debate over technology-related public policy issues; that government leaders are no more conversant with technology issues than the public; and that the lack of public interest may undermine the attraction of the engineering profession to young people. Thus, the engineering community cares, but the public should care also.

The engineering community already conducts a wide variety of public awareness activities. Available evidence indicates, however, that current activities have had no measurable impact on public awareness on a national scale. One purpose of this report is to provide a summary of a recent survey/questionnaire about public awareness activities. The questionnaire was commissioned by the National Academy of Engineering and conducted in late 2001 and early 2002. A second

TABLE 1-1 Greatest Engineering Achievements of the Twentieth Century

1. Electrification	11. Highways
2. Automobile	12. Spacecraft
3. Airplane	13. Internet
4. Water supply and distribution	14. Imaging
5. Electronics	15. Household appliances
6. Radio and television	16. Health technologies
7. Agricultural mechanization	17. Petroleum/petroleum technologies
8. Computers	18. Laser and fiber optics
9. Telephone	19. Nuclear technologies
10. Air conditioning and refrigeration	20. High-performance materials

Source: NAE, 2000.

purpose of the report is to recommend to the engineering community how public awareness activities should be refocused to have a national impact.

Engineers[1] have been integral to the designing and building of America, from the smallest microcircuit to the most massive civil infrastructure. Engineers translate scientific discoveries into the practical applications that maintain and improve our standard of living—food and water supplies, housing, electricity, sanitation systems, transportation, communications, security systems, medical devices and drug delivery, computers, and more.

The National Academy of Engineering, in cooperation with 27 professional engineering societies, recently compiled a list of the 20 greatest engineering achievements of the twentieth century (Table 1-1). This list was selected and ranked in order of importance by a committee of leading experts from academia and industry, and a wide range of engineering disciplines. These innovations, which shaped a century and changed the world, are used by or benefit the public every day with little awareness of how they happen to be available.

As a global economy emerges and global competition increases, it is important that the United States maintain its prominence and leadership in engineering, science, and technology. For that we will need a technologically sophisticated workforce. Our national security and competitiveness, as well as our standard of living, depend on our technology-driven industrial strength.

In the emerging interdependent global economy, sustained improvements in standards of living are expected, and in many places are desperately needed. The great challenges that lie ahead will demand the very best from the engineering

---

[1] There are 1.6 million engineers in the United States according to National Science Foundation data (NSF, 1999), based on the number of bachelor degrees, not including degrees in computer science, and 2.6 million engineers according to Bureau of Labor Statistics (BLS, 2002) data, based on occupational categories.

profession. The challenges will affect not only our country, but the entire world: competition for limited resources; global population growth; energy sourcing and security; homeland security; aging infrastructure in some parts of the world and the lack of infrastructure in others; water and air pollution; global warming; disposal of toxic waste; and vanishing habitats and endangered species. Great resourcefulness and ingenuity will be necessary to deal with these complex issues. The solutions that must be developed and implemented will require engineering innovations.

If the public as a whole and our leaders in particular are not able to make informed decisions about the uses of engineering innovations and technology, people can no longer oversee and ensure their welfare. Despite the personal and societal stakes, most people and political representatives at all levels of government do not have a sufficient understanding of how technology is developed and applied by the engineering professions. Currently, only eight members of the House of Representatives, one in the Senate, and a handful in state legislatures have engineering degrees. Only two members of the House of Representatives are licensed professional engineers.

Engineering is empowered to serve society by informed, aware citizens and policy makers. A well-informed public in a democratic society results in better policy and cost/benefit decisions. An aware public will seek information from and the participation of the engineering profession in important technological and standard-of-living issues, will understand the importance of engineering education and research and development (R&D), will make the changes necessary to improve them, and will encourage elected representatives to provide adequate support for them.

The public perception of the role of engineers is a major factor in whether or not talented young people consider engineering a desirable career choice. From 1900 to the middle 1980s, the number of graduating engineers with bachelor's degrees increased steadily. The number peaked in 1986 at 78,178 (see Appendix A), a healthy 7.9 percent of the total number of undergraduate degrees awarded that year. After that, the numbers declined to about 63,000 through the 1990s. In 2001, the number of bachelor's degrees awarded in engineering rebounded slightly to 65,113, comprising 5.5 percent of all B.S. degrees. Efforts to increase participation by women and underrepresented minorities from near zero levels have born fruit, but participation remains stubbornly low (NSB, 2000; see Appendix A).

In most of our key competitors in the industrial world, the number of engineers being graduated is substantially higher than in the United States. China produces three times the number of engineers, the European Union (EU) nearly twice as many, and Japan about two-thirds more (see Appendix A) (NSB, 2000).

Engineering education also provides a strong base for careers in other fields. Twenty-two percent of Fortune 200 CEOs have undergraduate engineering degrees, the most common degree held. Seventeen percent have liberal arts degrees, and nine percent have degrees in business administration (Neff and Ogden, 2002).

Although economic cycles may cause the total number of engineers needed by all industrial sectors to fluctuate somewhat, the underlying demand will continue to increase. For the United States to maintain its global technological lead and its standard of living, we must reinforce the value of young people pursuing an engineering education. An engineering-aware public would find ways to encourage, support, and reward its bright young people who seek careers in engineering.

REFERENCES

BLS (Bureau of Labor Statistics). 2002. Available online at www.bls.gov.

NAE (National Academy of Engineering). 2000. Greatest Engineering Achievements of the 20th Century. Available online at www.greatachievements.org.

Neff, T., and D. Ogden. 2002. Route to the top: the demand for performance. Chief Executive. Available online at www.chiefexecutive.net.

NSB (National Science Board). 2000. Science and Engineering Indicators–2000. NSB-00-1. Arlington, Va: National Science Foundation.

NSF (National Science Foundation). 1999. Scientists and Engineers Statistical Data System (SESTAT). Available online at http://srsstats.sbe.nsf.gov/.

2

Public Awareness Today

In 1998, the American Association of Engineering Societies (AAES) commissioned a survey of public awareness of engineering to be conducted by Harris Interactive. The survey was an attempt to provide better understanding of the results of a series of surveys conducted by Harris Interactive on a continuing basis on topics of interest, importance, and societal concern, such as the "prestige" Americans attach to various professions. The results of the earlier surveys are shown in Tables 2-1 and 2-2. Table 2-1 is a snapshot from 1998 that indicates that engineers have considerably less prestige than scientists. Table 2-2 shows that the percentage of respondents who consider engineers to have "very great" prestige has been "consistently mediocre" for 20+ years.

FOLLOW-ON SURVEY

In an effort to explore the results of the Harris polls, the AAES commissioned a follow-on study (also conducted by Harris Interactive). The results are analyzed below. In the first question, participants were asked to evaluate scientists, technicians, and engineers in the following terms:

1. When you hear the word engineer or scientist or technician, what first comes to mind about what that person does?
2. As some activities are mentioned, whom do you associate with that activity—a scientist, a technician, or an engineer?
3. As some characteristics are mentioned, who first comes to mind—a scientist, technician, or engineer?

TABLE 2-1 Excerpts from the Harris Poll Indicating the Level of Prestige American's Impart to Various Professions

	Very Great	Considerable	Some	Hardly Any	Don't Know
Doctor	61%	27%	10%	2%	1%
Scientist	55%	30%	10%	3%	1%
Teacher	53%	26%	15%	5%	1%
Minister	46%	28%	19%	7%	1%
Policeman	41%	31%	20%	7%	0%
Engineer	34%	39%	22%	4%	1%
Military Officer	34%	36%	23%	6%	1%
Architect	26%	42%	26%	4%	2%
Congressman	25%	31%	26%	17%	1%
Lawyer	23%	30%	28%	18%	1%
Athlete	20%	28%	34%	17%	0%
Entertainer	19%	29%	36%	15%	1%
Businessman	18%	37%	38%	6%	1%
Banker	18%	33%	39%	10%	0%
Accountant	17%	33%	38%	11%	1%
Journalist	15%	33%	37%	13%	1%
Union Leader	16%	28%	33%	22%	1%

Note: Not all of the percentages add up to 100 because not all respondents answered every question. *[The sample was a national cross-section of adults (N=1,000). The data are weighted to reflect the overall national population].*

TABLE 2-2 Results of Periodic Polls by Harris Interactive on the Prestige of Various Professions, 1977-1988
Percentage That Rated Prestige as "Very Great"

	1977	1982	1992	1997	1998
Doctor	61	55	50	52	61
Scientist	66	59	57	51	55
Teacher	29	28	41	49	53
Minister	41	42	38	45	46
Policeman	NA	NA	34	36	41
Engineer	34	30	37	32	34
Military Officer	NA	22	32	29	34
Architect	NA	NA	NA	NA	26
Congressman	NA	NA	24	23	25
Lawyer	36	30	25	19	23
Athlete	26	20	18	21	20
Artist	21	20	13	19	NA
Entertainer	18	16	17	18	19
Businessman	18	16	19	16	18
Banker	17	17	17	15	18
Accountant	NA	13	14	18	17
Union Leader	NA	NA	12	14	16
Journalist	17	16	15	15	15

The answers to these questions are shown in Table 2-3 and Figures 2-1 and 2-2.

TABLE 2-3 Comparative Perceptions of Scientists, Technicians, and Engineers

	Scientists	Technicians	Engineers
Invents	11%	2%	2%
Builds		10%	26%
Designs/plans	1%	1%	27%
Is creative	3%	1%	3%
Discovers	18%	1%	
Pioneers	1%		
Measures		1%	1%
Works in lab	8%	6%	
Conducts research	11%		
Cures diseases	9%		
Seeks knowledge	6%		
Conducts experiments	5%		
Equipment repair		15%	
Works w/ computers		9%	
Specially qualified in their field		6%	
Works w/ electronics	5%		
Train operator			5%

As Table 2-3 shows in response to question 1, the participants think of scientists as inventors and discoverers; of technicians as having specialized equipment-related qualifications; and of engineers as builders, makers, designers, and planners. The higher the respondent's educational level, the greater the likelihood that engineers were seen as designers and planners rather than builders.

Figure 2-1 shows the answers to question 2. Respondents strongly associated engineers with the design of new machines. They share credit with technicians for the development of software and with scientists and technicians for the design of medical instruments. However, most people did not recognize the contributions of engineers to the development of new forms of energy, to working in space, or to development of new drugs and medications. In general, scientists were more strongly associated with these activities.

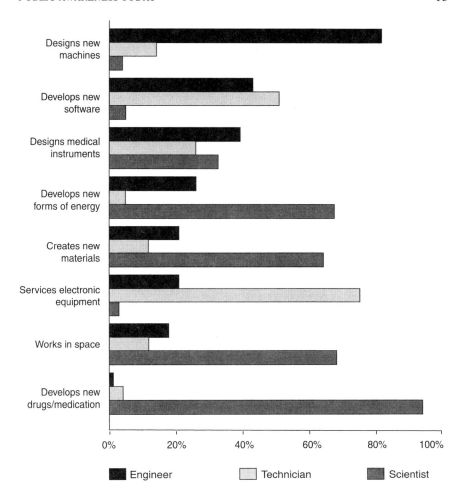

FIGURE 2-1 Activities Associated with Various Professions

Figure 2-2 shows the results of question 3. Engineers are perceived as more pragmatic contributors to society than technicians and less attuned to societal issues than scientists. The respondents credited engineers with contributions to economic growth, leadership, and national security, but gave them poor marks for contributions to our quality of life and the environment, for inclusiveness, and for social concerns.

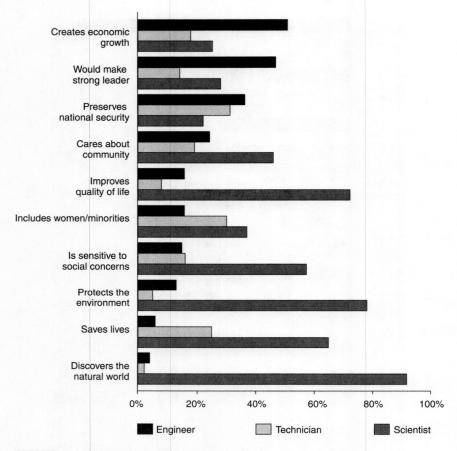

FIGURE 2-2 Characteristics Associated with Various Professions

Interviewees were then asked another set of questions:

4. Would you describe yourself as very well informed, fairly well informed, not very well informed, or not at all well informed about (a) science and scientists, (b) technology and technicians, (c) engineering and engineers?
5. Generally speaking, do you feel the media do an excellent, good, fair, or poor job in covering (a) science, (b) technology, (c) engineering, (d) medical discoveries?
6. If the media were rated "fair" or "poor," the subjects were asked if the reason was that the media *are not interested* in the discipline, or they think *you are not interested,* or *they don't understand* the discipline, or some other reason?

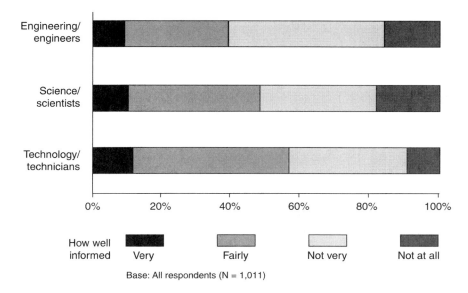

FIGURE 2-3 Level of Information about Science, Technology, and Engineering

Figure 2-3 shows that the respondents considered themselves not as well informed about engineers and engineering as about science and technology. A majority felt that they were "not very" or "not at all" well informed about engineering and engineers. The fraction of those who considered themselves poorly informed was somewhat higher among less educated people, and women considered themselves much less informed than men.

Figure 2-4 shows that, when the media are considered to do a "fair" or "poor" job covering engineering, the respondents believe the reason is that the media feel the public is not interested.

Respondents were then asked two final questions:

7. Do you feel that technology in general makes a positive contribution to society, makes a negative contribution to society, or doesn't have much impact one way or another?
8. Using a scale of 1 to 10, with 1 being extremely displeased and 10 being extremely pleased, if your son or daughter said they wanted to be a/an scientist/technician/engineer, how pleased would you be?

In answer to these questions, the respondents overwhelmingly (88%) felt that technology had made a positive contribution to society (Figure 2-5). Most respondents also said they would be very pleased (median source 9) for family members to become scientists, engineers, or technicians. Because the profes-

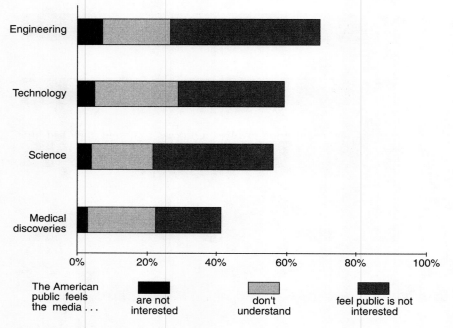

FIGURE 2-4 Media Coverage

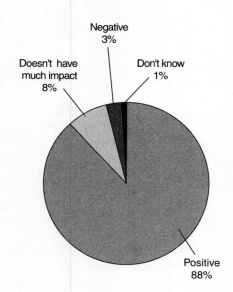

FIGURE 2-5 Contribution of Science, Technology, and Engineering

sions were grouped together in question 8, it was impossible to separate out engineering.

CONCLUSION

The results of the periodic Harris Surveys and the follow-on survey commissioned by AAES paint a disappointing picture of the public understanding of engineering and public perceptions of the profession. Although most respondents recognized that engineering involves a process of design, they had little sense that engineering also involves applications of those designs. They expressed a great deal of goodwill toward technology, but seemed to direct only a modicum of that goodwill toward engineers and engineering. Although the collective efforts of engineers dramatically improved the quality of life for Americans in the twentieth century, the public appeared to give them little credit for those contributions. The survey results in Chapter 3 indicate the engineering community has been conducting a variety of public-awareness and outreach programs for many years, some reportedly for more than 20 years. Although some of these programs are well known locally, the Harris Surveys indicate that, on a national basis, they have not improved the public's perception of engineers and engineering. Perhaps, at best, they have forestalled deterioration of public opinion.

3

The National Academy of Engineering Survey of Public Awareness of Engineering Outreach Programs

In April 2001, The National Academy of Engineering (NAE), with funding from the Elizabeth and Stephen Bechtel, Jr., Foundation, initiated a new project as part of its Public Understanding of Engineering Program. The project involves conducting a survey/questionnaire and creating an inventory of current outreach programs for improving public awareness of engineering and developing recommendations based on the survey results. The recommendations will also suggest other activities to increase the public awareness of engineering. To carry out the project, NAE formed the Committee on Public Awareness of Engineering (CPAE) composed of a group of distinguished citizens interested in and/or involved in the engineering community (Appendix D). The committee reviewed the results of the survey/questionnaire and made recommendations for the next phase of the program.

The survey/questionnaire was developed jointly by NAE and outside consultants, Market Research Bureau LLC (MRB) and McMahon Communications. The consultants administered the survey, tabulated the responses, and provided an analysis of the results. They polled a broad spectrum of organizations involved in engineering to determine what they or their organizations were doing to improve the public perception and awareness of engineering and to determine "best practices" (i.e., the most effective programs and/or techniques).

The study adopted the following goals:

- Identify best practices in current programs to improve the public understanding and appreciation of the role of engineering in society.
- Recommend ways the engineering community might leverage these best practices.

- Formulate consistent core message(s) (e.g., slogans, catchwords, etc.) for all future outreach activities.
- Recommend additional ways the engineering community might improve its individual and collective efforts to increase the public awareness of engineering.
- Suggest to NAE and the engineering community a focus and activities for the next phase of this initiative.

Appendix D lists the membership of the CPAE, the NAE staff involved, and the members of an Advisory Group that met with NAE staff and consultants to help shape the suggested recommendation for the CPAE to consider.

OBJECTIVES AND METHODOLOGY

The main purpose of this project is "to help the engineering community maximize its resources to deliver a comprehensive, coordinated, and sustained message that will help the public better appreciate the fundamental importance of engineering to the quality of their lives and to the productivity and economic strength of the nation. The findings of the study will be used as a foundation from which the engineering community can work together to impact public perceptions, public policy decisions, and the education system."

The study identifies and analyzes the scope, nature, objectives, and effectiveness of current engineering communications, education, and outreach activities by various engineering organizations. The committee attempts to identify gaps or deficiencies in current activities and recommend ways to develop consistent messages and strategies for a coordinated effort.

The data for the study were collected via a self-administered questionnaire, (Appendix E) that was sent to 628 organizations. The potential respondents could be divided into 10 broad categories covering all aspects of the engineering community. Specific individuals for each type organization were asked to complete the questionnaires: executive directors of engineering societies; executive directors of industry associations; executive directors of educational associations; directors or CEOs of museums; directors of national laboratories; CEOs or chief technology officers of private companies; CEOs of design/contracting firms; deans of engineering schools at colleges and universities; midlevel or senior-level administrators at federal agencies; and senior-level producers at media outlets. Lists of engineering societies, museums, national laboratories, federal agencies, and producers/media lists were provided by NAE. The other lists were put together by MRB and McMahon Communications in conjunction with NAE. The industry list includes appropriate companies on the 2001 Fortune 500 list. The design/contracting firms include the top 50 design and/or contracting firms as ranked by the *Engineering News Record*. Colleges and universities were selected using various rankings of the top engineering schools (*US News and World*

Report, Peterson's Undergraduate Guide: Four-Year Colleges 2002, and *The Best Graduate Programs: Engineering*). A list of responding organizations is given in Appendix F.

Once respondent organizations had been identified, Stephen D. Bechtel, Jr., CPAE Chair, sent a letter to each potential participant explaining the purpose and importance of developing an inventory of current activities advising them that they would be receiving the questionnaire. The initial letters were mailed on November 9, 2001. The questionnaire was accompanied by another letter from Mr. Bechtel on NAE stationary. A postage-paid envelope was included for the return of the questionnaire to Market Research Bureau. The questionnaire was also posted on the NAE web site and could be downloaded and completed by hand or electronically. Sixty-seven respondents completed the questionnaire electronically. The questionnaires were mailed on November 16, 2001, and respondents were given a due date of December 7, 2001. Letters and questionnaires to the museums were mailed approximately two weeks after the original mailing dates, and their due date was extended accordingly to December 21, 2001. In the event, however, "late" questionnaires were accepted through February.

From late December 2001 through January 2002, individuals who had not returned the questionnaires were contacted by telephone. Some respondents indicated that they did not have any outreach activities, and a few agreed to complete the questionnaire by phone. Other organizations were contacted by NAE or Mr. Bechtel's office, which yielded several more responses. The overall response rate for the study was 39 percent with variations by respondent category (Table 3-1).

In most categories, the questionnaires were completed by the addressees. In the industry and design/construction firms categories, however, the questionnaires were passed along to a communications department, a Chief Technology Officer (CTO), or an engineering department. Responses were confidential, and respondents will be provided with the results of the study. Most of the questionnaires were returned directly to MRB either by mail or electronically; a few were sent to NAE and then forwarded to MRB.

The overall response rate is excellent, especially in light of the length of the questionnaire, the high caliber of the respondents, and the absence of any incentive other than the importance of the study itself. The high response rate can probably be attributed to the perceived importance of the study, the respondents' pride in their activities, NAE's sponsorship of the study, and Mr. Bechtel's endorsement of the study. The typical response rate for self-administered questionnaires among a professional audience is much lower.

There is probably a response bias in that individuals whose organizations have activities were more likely to return the questionnaire than those whose organizations do not have any activities. This probably accounts for the differences in the response rate by category. For example, engineering societies, colleges and universities, and national laboratories, which had the highest response

TABLE 3-1 Questionnaire Response Rates

	Number Sent	Undeliverable	Questionnaire Returned	No Outreach[a]	Return Rate[b] (percent)	Expressly Refused/ Declined
Engineering Societies	80		50	6	70.0	1
Industry	237	3	54	12	28.2	21
Design and Contracting Firms	47		12		25.5	2
Colleges and Universities	133		60		45.1	2
Industry Associations	17		7		41.2	
Educational Associations	16		7		43.8	1
National Laboratories	22		12	1	59.1	
Federal Contacts	17		5	2	41.2	
Producers/Media	22		4		18.2	
Museums	37		13		35.1	
TOTAL	628	3	224	21	39.2	27

[a] Only respondents who indicated that they did not return the questionnaire because they had no outreach activities. Respondents who indicated they had no outreach activities on returned questionnaires are included in the "questionnaire returned" count.
[b] The response rate was calculated as follows: (returned + no outreach)/(sent − undeliverable).

rate are more likely to consider engineering outreach or education as a part of their mission. Although the study may not be all inclusive, it most likely represents the vast majority of current engineering outreach activities and should be considered a representative assessment of the current situation. Specific smaller programs that may be missing are likely to be similar to the programs that are represented, and their absence does not change the overall conclusions of the study.

The questionnaire is a comprehensive, 12-page document covering all possible activities. Although the questionnaire was structured, respondents were encouraged to add comments, and many did. Respondents were also asked to include samples of materials, and about one-third provided samples.

In addition to the self-administered questionnaire, a number of in-depth interviews were conducted to clarify responses or to obtain additional information from respondents whose organizations had programs that were particularly interesting. Because both objective and subjective data were collected through interviews and extensive responses to open-ended questions, the data were analyzed in both quantitative and qualitative terms. The analysis is presented in the remainder of this chapter. A section highlighting selected outreach programs can be found in Appendix C.

THE CURRENT SITUATION

Current Activities

There is a great deal of outreach, communications, and educational activity being done by the engineering community. Of the 245 organizations (224 that returned questionnaires plus 21 that indicated they had no outreach activities) that responded in some way to this study, 72 percent indicated that they are currently engaged in some sort of communication, outreach, or educational activity (Table 3-2).

Although all types of organizations have outreach activities, colleges and universities, engineering societies, museums, and national laboratories are most likely to consider outreach as a part of their mission. The mission statements of many educational institutions and societies include promoting the engineering profession through outreach activities. The focus of activities by museums is on general science and technology rather than specifically on engineering. Private companies, both general industrial firms and engineering firms, are most likely to expect direct benefits from their efforts rather than long-term improvements in the engineering profession. These companies need engineers to produce their products and are primarily trying to encourage young people to go into the field. The purpose of media/producers activities is to provide education but also to provide entertainment; their programs are not structured to produce a long-term effect on engineering. Some sample mission statements for different types of organizations are provided in Appendix B.

Most current activities are ongoing programs; many sponsoring organizations have both ongoing and discrete programs (Table 3-3).

The majority of organizations with outreach activities have multiple programs in place and have been doing this type of activity for many years (Table 3-4).

Most organizations have been conducting activities for more than 10 years. Twenty percent indicated that their programs have been ongoing for 25 years or more (Table 3-5).

TABLE 3-2 Is your organization currently engaged in communications, educational, or outreach activities designed to improve the public understanding of engineering?

	Percentage
Yes	72
No	26
No answer	2

Total of respondents who returned the questionnaire or indicated they have no outreach activities = 245.

TABLE 3-3 Are these ongoing activities or discrete programs with limited duration?

	Percentage
Ongoing only	40
Discrete only	5
Both ongoing and discrete	50
No answer	5

Total number of respondents with activities/programs = 177.

TABLE 3-4 Does your organization have more than one program of such activities (i.e., that reaches different audiences, has different objectives or different messages)?

	Percentage
Only one	9
More than one	86
No answer	5

Total number of respondents with activities/programs = 177.
Note: The highest number reported was 300; among those who provided a numerical response, the median was three.

TABLE 3-5 For how long has your organization had some sort of communications, education, or outreach effort?

	Percentage
Less than 3 years	3
3 to 5 years	6
5 to 7 years	4
7 to 10 years	6
More than 10 years	73
No answer	8

Total number of respondents with activities/programs = 177.

Most respondents indicated that their organizations are committed to continuing their activities for the foreseeable future. Those who currently had plans for the future were asked if they had plans for new programs, as well as (or in place of), current programs (Table 3-6).

TABLE 3-6 Does your organization have plans for any communications, education or outreach activities in the future (beyond what you might be doing right now)? If you are doing something currently, do you have any plans for *other* efforts once the current activities are completed?

	Percentage
Have plans [a]	56
Do not have plans	36
No answer	9

Total number of respondents with activities/programs = 177.
[a] Includes continuing current programs and new programs. About 10 percent indicated that they were referring only to ongoing programs.

The purposes of current activities vary, but a substantial number of programs are targeted toward young people (K-12). Many programs attempt to introduce young people to engineering or to stress the importance of math and science. The messages conveyed by these programs are primarily that math and science are fun and that mastering them can lead to rewarding, challenging, fun, exciting, and creative careers in engineering.

A number of respondents listed the general public as a target audience. The descriptions of the programs, however, indicated that they were focused more on educating or, more accurately, inspiring young people to learn about engineering, technology, or math and science. The main concern of colleges and universities is increasing enrollments and filling the "pipeline" of engineers for industry; industry is also concerned about the pipeline issue. Concerns about general perceptions of engineering were slanted toward their impact on the pipeline issue.

Eighty-two percent of respondents with outreach programs indicated that their programs targeted children in K-12, especially older children (grades 9-12.) Some highly visible programs are national in scope (most notably National Engineers Week, although it is typically conducted as a series of local projects), but most activities are local in scope (e.g., mentoring programs, competitions, speakers, etc., at local schools or community venues) (Table 3-7).

The messages conveyed by most programs can be characterized as recruitment messages: "engineering is a fun, creative, exciting, important career"; "math and science are fun" messages for younger children; and "engineers are important and contribute to the quality of life, the economy, and environment" messages. These general messages appeared over and over again, although the wording differed from program to program.

Organizations use a variety of tools in their programs; the most common are web sites, public relations, speakers, and informal educational programs. Most organizations use more than one approach (Table 3-8).

TABLE 3-7 What are the target audiences for your organization's communications, educational or outreach efforts? Who are you trying to reach with your messages?

	Percentage
General public	69
Engineers	60
Potential clients/engineering users	49
Undergraduate students	62
Children in kindergarten through fifth grade (K-5)	43
Children in sixth through eighth grade (6-8)	62
Children in ninth through twelfth grades (9-12)	77
Unduplicated K-12	82
Teachers in kindergarten through twelfth grades (K-12)	59
College/university faculty	59
Opinion leaders	49
Public policy makers	49
Newspapers	53
Broadcast media	49
Other media	19
Other audience	18

Total number of respondents with activities/programs = 177.
This table shows the percentage of respondent organizations that address each of these audiences but does not show the percentage of programs that address each audience.

TABLE 3-8 Which of the following specific activities is your organization using in its communications or outreach efforts?

	Percentage
Web sites	77
Speakers/symposia	66
Public relations	65
Informal educational programs	57
Direct mail	46
Formal educational curriculum	38
Public affairs/policy	37
Paid advertising	35
Public service advertising	19
800 number	16
Other	19

Total number of respondents with activities/programs = 177.

When respondents were asked to answer questions pertaining to the specific activities they engage in, more respondents answered some specific questions than might be expected based on the general responses in Table 3-8. To capture the greatest amount of information, the responses to subsequent questions are tabulated based on actual responses.

Web Sites

The Internet is the most widely used tool, but in most cases it is an extension of the main web site and not a separate site for outreach activities (Table 3-9).

A little more than one-third (39 percent) have separate links on their web sites for the media (Table 3-10).

Speakers/Symposia/Forums

Speakers and other in-person activities are used by about two-thirds of the respondents. About one-third have speakers on staff (Table 3-11, Table 3-12).

Public Relations

Respondents whose organizations engaged in public relations were also asked to indicate which media they targeted. Once again, print media (e.g., newspapers and trade/professional journals) were cited most often (Table 3-13).

Although about two-thirds of those who engage in outreach activities use public relations, only about one-third actually have press kits (Table 3-14).

About half have met with the media to discuss their programs (Table 3-15).

TABLE 3-9 Is the web site your organization uses in its communications effort specifically designed for that effort or is the site also used for other reasons (e.g., member information)? If the same, is there a separate link or icon on the site directed at the target audience for your communications effort?

	Percentage
Same web site	79
Separate link	53
No separate link	26
Different web site	7
No answer	14

Total number of respondents with activities/programs = 177.

TABLE 3-10 Is there a link or icon for the media or press on your organization's web site?

	Percentage
Separate link for media	39
No separate link for media	44
No answer	17

Total number of respondents with activities/programs = 177.

TABLE 3-11 Do you participate in and/or sponsor speakers/symposia/forums?

	Speakers (%)	Symposia (%)	Forums (%)
Participate only	14	6	6
Sponsor only	2	3	2
Both participate and sponsor	49	38	35
No answer	35	52	58

Total number of respondents with activities/programs = 177.

TABLE 3-12 Do you have speakers on staff?

	Percentage
Yes	33
No	45
No answer	21

Total number of respondents with activities/programs = 177.

TABLE 3-13 Do you target any particular media?

	Percentage
Newspapers	58
Trade/professional publications	51
Web-based media	40
Television	39
Radio	36
Consumer magazines	17
Other	6

Total number of respondents with activities/programs = 177.

TABLE 3-14 Do you have a press kit?

	Percentage
Yes	35
No	43
No answer	22

Total number of respondents with activities/programs = 177.

TABLE 3-15 Have you scheduled any meeting(s) with a news organization(s) in the past year to explain the work of your organization or institution?

	Percentage
Yes	48
No	33
No answer	19

Total number of respondents with activities/programs = 177.

TABLE 3-16 What grades are your programs designed for?

	Percentage
K-5	38
6-8	53
9-12	68
College	57
Graduate school	43
Adult	38

Total number of respondents with activities/programs = 177.

Educational Programs for Engineering & Technology

Educational programs are mostly targeted at older children—especially high schoolers and undergraduates (Table 3-16).

National Engineers Week, competitions, and mentor programs are the most often mentioned educational activities (Table 3-17).

Most of the organizations with educational outreach programs have developed their own materials (Table 3-18).

About two-thirds of educational activities are designed specifically to promote engineering; one-third promote a general science curriculum (Table 3-19).

TABLE 3-17 Which of the following activities do you provide?

	Percentage
National Engineers Week	49
Competitions	48
Mentor programs	46
Formal curriculum	36
Materials to guidance counselors	32
School-to-work training	18
Other	36

Total number of respondents with activities/programs = 177.

TABLE 3-18 Did your organization develop the curriculum/materials or did you get it/them someplace else?

	Percentage
Developed by respondent	51
Developed elsewhere	10
Both	16
No answer	23

Total number of respondents with activities/programs = 177.

TABLE 3-19 Are the educational materials you use specifically aimed at educating students about engineering and technology distinct from science?

	Percentage
Yes, designed for engineering	42
No, part of science	32
Both	3
No answer	23

Total number of respondents with activities/programs = 177.

Public Affairs/Public Policy

Although only 37 percent of respondents indicated that they engaged in some kind of public policy activity in the general question (Table 3-8), almost 60 percent indicated that they engage in lobbying or briefings of some sort. Almost two-thirds (63 percent) indicated that they had met with elected officials. The inconsistency may be attributable to terminology (Table 3-20, Table 3-21).

TABLE 3-20 What type of public policy or public affairs activities does your organization engage in?

	Percentage
Briefings	44
Lobbying	34
Other	18
None/no answer	41

Total number of respondents with activities/programs = 177.

TABLE 3-21 Have you had meetings with elected officials to explain your work and/or resources?

	Percentage
Yes	63
No	18
No answer	20

Total number of respondents with activities/programs = 177.

Paid or Public Service Advertising

Respondents whose organizations use paid or public service advertising were asked which media they use. The Internet (although there might be some confusion between having a web site and actually advertising on the web) and print media (primarily trade and professional publications, but also newspapers) were mentioned most often. Local television and radio were also mentioned by a fair number of respondents (Table 3-22).

Advertising budgets (of the few that were provided) ranged from $1,000 to $15 million (including a production budget of $4 million). The call to action cited most was "Visit our web site."

Monitoring/Effectiveness

The vast majority of respondents believe their programs are successful, mostly based on measurements of short-term processes or tactics (e.g., number of participants, number of web site hits, or number of media placements) rather than longer-term outcomes (Table 3-23, Table 3-24).

TABLE 3-22 What media do you use?

	Percentage
Internet	37
Trade/professional publications	35
Newspapers	35
Local/spot radio	25
Local/spot TV	22
Local cable TV	14
Network broadcast TV	10
Outdoor	10
National cable TV	9
Network radio	9
Consumer magazines	9
Other	8

Total number of respondents with activities/programs = 177.

TABLE 3-23 What criteria are important in determining if your communications effort is successful?

	Percentage
Awareness	75
Knowledge	69
Attitudes	60
Change in behavior	48
Other	13

Total number of respondents with activities/programs = 177.

TABLE 3-24 Which, if any, of the following do you use in evaluating your programs?

	Percentage
Web site hits	51
News coverage in print, broadcast, and/or online media	48
Exit interviews or on-site evaluations (for speakers, symposia, forums)	36
Tracking of research (market research to track awareness or attitudes)	22
Response tracking (for direct mail or direct response advertising)	20
Placement reports (for public service advertising)	9
Other	17

Total number of respondents with activities/programs = 177.

Few respondents felt that their programs were not successful. Those who did attributed the lack of success to a lack of adequate resources (e.g., time, staffing, money). The positive self-assessments, combined with the willingness of respondents to share the details of their programs, indicates that organizations have a great deal of pride in their outreach activities (Table 3-25).

In the absence of measures of long-term effectiveness, it was impossible to meet a major goal of the survey — to identify "best practices." See Appendix C for a list of programs thought by the CPAE to be particularly interesting and well constructed.

Planning and Decision Making

Most organizations have separate departments dedicated to outreach activities, but they are usually small; 54 percent indicated that their departments had six people or fewer (Table 3-26).

Respondents were asked about their total budgets for communications and outreach. Some were reluctant to provide information about budgets, but among those who did, budgets ranged from zero to more than $50 million; most of the higher figures included staffing.

TABLE 3-25 Do you feel your efforts have been successful in terms of meeting the original objectives?

	Percentage[a]
Successful	79
Unsuccessful	11
Don't know/too early to tell	10

Total number of respondents with activities/programs = 177.
[a]This table indicates the percentage of respondent organizations that considered any of their programs successful or unsuccessful. It does not show the percentage of all programs that address each audience.

TABLE 3-26 Does your organization have a department dedicated to communication, education, or outreach?

	Percentage
Have department	65
Do not have department	29
No answer	6

Total number of respondents with activities/programs = 177.

The total expenditures for all organizations that provided budget information is $264,438,587. To estimate the total expenditures for all respondents, we multiplied the mean budget of those that provided information by the total number of respondents (n=177). The total estimate is approximately $403 million. These numbers do not include the value of the time of the many engineers who assist outreach programs as volunteers. Decisions, including budget decisions, are usually made by more than one person; in most cases, management of the organization is involved in outreach decisions.

Most responding organizations did not have written communications plans. Even organizations with substantial budgets did not all have written plans (Table 3-27).

Other Programs

Most respondents (53 percent of all respondents to the questionnaire) indicated that they were aware of other outreach programs (Table 3-28). Not surprisingly, those who had their own programs were more likely to be aware of other programs than those who were not engaged in outreach (59 percent versus 32 percent).

The programs most often mentioned were sponsored by NAE, National Society of Professional Engineers (NSPE) (while National Engineers Week is now a separate non-profit organization sponsored by engineering societies and industry on a rotating basis, many respondents still identify it with NSPE, due to their long stewardship of the program), American Society of Civil Engineers, National Science Foundation, Institute of Electrical and Electronics Engineers, Inc., and American Society of Mechanical Engineers. Many respondents simply referred to "other professional societies" or "other colleges or universities." It should be noted that throughout the questionnaire National Engineers Week was referred to by various names [e.g., Engineers Week, Engineering Week, NEW, E-Week.]

No single program was recognized as particularly effective. National Engi-

TABLE 3–27 Do you have a written public communications, education or outreach plan?

	Percentage
Yes	32
No	54
No answer	14

Total number of respondents with activities/programs =177.

TABLE 3-28 Are you aware of any other organizations that are engaging in communications, education, or outreach programs on engineering?

	Percentage
Yes, aware	53
Not aware	33
No answer	14

Total number of respondents = 224.

TABLE 3-29 Are you presently partnering or coordinating with any other organization for any of your current or most recent communications or outreach efforts?

	Percentage
Partnering	44
Not partnering	40
No answer	16

Total number of respondents = 224.

TABLE 3-30 Have you ever partnered or coordinated with any other organization in any of your communications efforts?

	Percentage
Have partnered	40
Have not partnered	37
No answer	23

Total number of respondents = 224.

neers Week was the most visible and entrenched. When asked which programs they considered most effective, most respondents did not name any.

Many organizations have experience in partnering with other organizations (Table 3-29, Table 3-30). These partnerships range from working with other departments within their organizations to using materials supplied by others to working directly with schools, community groups, and other organizations to develop and distribute materials and messages. The most commonly cited partnering organizations are shown in Table 3-31.

TABLE 3-31 Most Commonly Cited Partnering Organizations

Organization	Number of Citations
NSPE/National Engineers Week	23
AAES	8
ASCE	6
NAE	5
NSF	4

TABLE 3-32 Do you feel there is a need for coordinated communications efforts that would encourage consistent messages and possibly provide efficiency for various organizations interested in communicating about engineering?

	Percentage
Yes	73
No	10
Don't know/maybe	4
No answer	13

Total number of respondents = 224.

Coordinated Efforts

The questionnaire revealed strong, but not universal, support for a coordinated program (Table 3-32). Most respondents recognize that a coordinated program would be more efficient and more effective than individual programs because it would deliver a consistent message. When asked what that message should be, most suggested that it should communicate the importance of engineers to society. Many also suggested a message about the value and satisfaction of a career in engineering.

When asked why a coordinated effort is needed, respondents gave several reasons:

- No one organization "speaks" for the engineering community as a whole. Different organizations have different priorities, different goals, and different messages. Engineers are not public relations oriented.
- Leveraging dollars and telling a good story can best be done through a coordinated program.
- The engineering community needs to get a coherent message out.

- The public does not know who engineers are or what they do. A coordinated effort to increase awareness of the industry would benefit the entire community.
- Society as a whole needs to have a better understanding of the role and importance of engineering and to appreciate the value and excitement of an engineering career. Engineers are a diverse group of "normal," dedicated individuals. The engineering discipline is not narrow or limited in scope.
- The profession needs better visibility. A coordinated program would bring deserved recognition of engineers as professionals and of the valuable work they do to make our world better. This might lead to a more equitable fee structure for engineering consulting work.
- Providing information to a large audience is expensive, few associations have the resources to do this on their own.

Most of those (minority) who do not believe a coordinated effort is necessary cited the difficulty of administering such a large program rather than citing a reason it was not needed. Typical answers are listed below:

- We must be careful to avoid consuming limited human resources for cooperative, coordinated efforts that are not effective.
- Engineering fields are too diverse to be combined in a single program.
- A coordinated, large-scale program would be too complicated to administer efficiently.
- Messages for many engineering organizations are too diverse to be coordinated.
- Most organizations operate with organization-specific goals and objectives.
- It will be futile to try to "educate" the general public about engineering, like trying to teach a rock to talk. It would not be worth the effort because the public does not *care*.
- People do appreciate engineers but we must encourage children to enter the sciences.

Most respondents also said they were prepared to participate, or at least consider participating, in a coordinated program. A few qualified their comments by indicating that a coordinated effort should complement, rather than replace their programs. They also indicated that their resources (e.g., money and staffing) are committed to these existing programs and are already stretched too thin (Table 3-33).

TABLE 3-33 Would your organization be interested in participating in such co-ordinated communications efforts?

	Percentage
Yes	67
No	16
Don't know/maybe	3
No answer	14

Total number of respondents = 224.

If we break down the answers to questions about a coordinated effort and the willingness to participate by the type of organization, some differences can be noted. Colleges and engineering societies are more likely to see a need for, and are more willing to participate in, a coordinated effort than industrial companies. Percentages for the categories with sufficient bases to be significant are presented in Table 3-34a. The number of respondents is shown for other categories (Table 3-34b).

TABLE 3-34a Do you feel there is a need for a coordinated effort?

	Engineering Societies in percent	Industry in percent	Colleges/ Universities in percent	Total in percent
Is there a need?				
Yes	76	71	78	73
No	8	16	12	10
Don't know	4	—	—	4
No answer	12	13	10	13
Willing to participate?				
Yes	78	53	75	67
No	8	29	10	16
Don't know	—	7	—	3
No answer	14	11	15	14
Totals	(50)	(54)	(60)	(224)

Base-Total respondents in each category as indicated

TABLE 3-34b Actual numbers of respondents rather than percentages.

	Design/ Contracting	Industry Associations	Education Associations	National Laboratories	Federal Contracts	Producers/ Media	Museums
Is there a need?							
Yes	9	3	4	9	4	4	6
No	—	1	—	—	—	—	2
Don't know	1	1	2	1	—	—	2
No answer	2	2	1	2	1	—	3
Willing to participate?							
Yes	7	1	4	8	4	4	9
No	1	3	3	—	1	—	1
Don't know	2	—	—	1	—	—	—
No answer	2	3	—	3	—	—	3
Totals	(12)	(7)	(7)	(12)	(5)	(4)	(13)

Base-Total respondents in each category as indicated

INTERPRETATION OF FINDINGS

This interpretation is based on the quantitative findings and comments on the questionnaires provided by respondents, secondary research and one-on-one interviews with respondents by NAE's consultants.

The Effectiveness of Current Outreach Activities

Conclusion 1. Grassroots involvement in outreach programs is widespread and has had some benefits. Nevertheless, these programs have not had a demonstrated impact on enrollments in engineering programs or improved the public awareness of engineering.

One of the most striking findings of the NAE inventory questionnaire is the widespread awareness at the grassroots level of the negative effects of declining literacy in math and science and declining engineering enrollments in engineering on U.S. competitiveness, national security, and standard of living. Some engineers noted that higher visibility and appreciation would be gratifying, but they ultimately recognized that more serious issues must be addressed.

To address those issues, many programs have been initiated to reach students and spark an interest in engineering. Competitions in building bridges, cars, and robotic systems abound; some scouting programs are focused on engineering; math competitions are legion; National Engineers Week continues to attract widespread interest; even "camps" are focused on math and engineering. These and many other programs are based on the same ideas—showing students the applicability of math and science and enabling them to interact with engineers in small groups. All of these programs rely on engineers volunteering to work with students in extracurricular activities and visiting classrooms to talk about their careers.

Clearly, many engineering firms and corporations respond affirmatively when they are asked by a national or local organization to donate their time and money. Many also participate in National Engineers Week, during which engineers visit classrooms as part of a national program. Participants in these programs are uniformly enthusiastic about them and report that they enjoy interacting with students. For many firms and businesses, the underlying principle behind their participation is visibility for their organizations and an opportunity for recruiting. Several respondents noted that they track, either formally or informally, the numbers of students who later intern at their organizations, pursue engineering degrees, and return to their companies to work. When asked about the need for a coordinated effort to solve the problems inherent in attracting engineering students, several said they would not want to give up the activities they already support to take on something new, partly because of the importance of these activities in recruiting. In many communities, these activities have been entrenched for many years, and the participants are part of a community "network" that makes the activities possible.

Although there are a large number of grassroots activities, they have no clear effect on enrollments or increased math literacy; and, in fairness, they are not designed to do so. Contacts between engineers and students are intermittent, because many of these activities take place only once a year. The students are largely self-selected, who choose to spend their free time in these activities on an "opt-in" basis; for that reason, the programs generally appeal to students who have already shown an interest in, and an aptitude for, math. By their nature, these programs do not reach students across the spectrum of gender, race, and ethnicity, because they are available only to select schools and communities. As one respondent said, "We're talking to our sons and daughters" in these programs, suggesting that many of the students who take part in these programs would find their way to math class and engineering school anyway. Most of their parents and teachers are involved and well-educated enough to steer them toward extracurricular programs that give them a chance to broaden their skills and understanding.

Conclusion 2. There is a growing realization that successful outreach to students must begin at the K-3 level.

The responses to the questionnaire indicate that outreach programs are skewed toward upper grade levels. The programs mentioned above, either classroom visits as part of National Engineers Week (E-Week) or participation in extracurricular activities, are designed for students in the upper grades for obvious reasons. As math classes become more advanced, students begin to self-select out of them. The remaining students are the ones who are capable of taking part in these academic and hands-on competitions and programs.

Several respondents noted that little is being done at the elementary level even though those are the years when the foundation is laid for liking math (and science) and for believing in the possibilities of success. One respondent who volunteers in classrooms as part of her company's outreach program noted that children at the K-3 level routinely consider science and math classes as fun. By the sixth grade, however, these classes are the least favorite, and many students believe that they will not do well in these subjects. One university engineering school that teamed up with educational professionals to conduct research into math and science learning concluded that the beginnings of career pathways are in elementary school when children decide which subjects they like and will be good at. Their studies show that children who are turned off to math and science between grades 2 and 4 have made a permanent decision. On that basis, they concluded that future recruiting efforts may depend on the ability of science and engineering professionals to influence school children at an early age (Mathias-Riegel, 2001).

Many other respondents who did not have academic credentials or the benefit of research voiced the same opinion based on their years of volunteering with students and visiting classrooms. Repeatedly, respondents said that encouraging middle school and high school students to consider engineering careers is not effective, either because they have already made up their minds to pursue other interests, or because they are afraid they could not compete successfully in an engineering school or because they had opted out of advanced math classes early and are not prepared to pursue an engineering track.

Conclusion 3. Many engineering schools are actively looking for ways to increase student enrollments; several are involved in sophisticated and creative current outreach programs.

Several engineering schools have undertaken impressive outreach programs and have engaged in off-campus partnerships that not only provide immediate benefits to their cities and states, but also have the potential to make long-term, fundamental changes. In three separate settings, two on the East Coast and one in the Midwest, engineering schools have formed partnerships with state and local institutions to expand their reach and strengthen their education offerings. One engineering school has established a successful summer camp for math students, which is significantly underwritten by the state business and corporate community. A second engineering school has formed an alliance with the state department of education and other institutions to improve the K-12 curriculum and to train current teachers to teach to those standards. A third engineering school has entered into a regional partnership with eight other universities and the public school system in its metropolitan region to bring hundreds of students, many of them minority students, to their campuses on a year-round basis. College faculty coach the students in the math, science, and computer skills that they will need to succeed in college. In this program, the business and corporate community supplies the funds to pay the faculty for their teaching time.

Conclusion 4. Many people in the engineering community advocate more visibility for engineers in television and movies. The media offer both opportunities and risks.

Engineers are largely absent in popular culture, which is the way many kids—and many adults—receive information. Many respondents noted that movies and television shows rarely feature characters who earn their living as engineers or plots that revolve around engineering firms or situations. There was a general feeling among respondents that just a few references to engineers on prime time TV and in feature-length movies would go a long way towards raising awareness.

We feel obliged to repeat the adage, "Be careful what you wish for." Prime time series that feature doctors, lawyers, emergency medical teams, and police officers rely on the truism that people who work in these jobs routinely help good people who have been hurt by situations not of their making, typically in crisis situations. Although engineers are sometimes in that position, securing a rescue site, for example, so that rescue workers can enter it safely, they are more often involved in the long-term design and planning of projects like office buildings, transportation systems, water and wastewater systems, agricultural systems, and medical equipment. To making these exciting enough for prime time viewers, plot twists could go something like this: an incompetent, unethical engineering firm on the take designs a bridge that collapses, trapping and injuring small children in school buses on the span; competent engineers secure the site for rescue workers, help carry the kids to safety, and blow the whistle on the "bad guy" engineers.

Prime time shows about doctors and lawyers routinely portray lives that are ruined because of inept doctors and vicious lawyers. Nevertheless, real-life doctors, at least, consistently score high in the trust and respect category. Engineers, however, although respected by the public, may not have a reservoir of goodwill to sustain them, and would hardly enjoy being portrayed as vicious, incompetent, uncaring, or crooked. Television networks should be approached very carefully, with an understanding of the trade-offs between accurate portrayals and series ratings.

An alternative idea is to mount a concerted campaign to influence Saturday morning programming for youngsters. Sophisticated cartoons might lend themselves to an engineering character, plot, or references.

Feature length movies may represent a real opportunity. It is suggested that the approach be CEO to CEO. For instance, a CEO-level meeting at Disney and Pixar (which teamed to make the very popular "Toy Story") would be one avenue of pursuit; the case could be made that a positive portrayal of engineers could encourage youngsters in that direction. CEOs of engineering companies would explain the long-term implications of declining engineering enrollments, stressing the business issues of competitiveness, standard of living, and national security, and ask the studio to consider adding an engineering character to a movie already in development or writing an engineering character into a new script. The emphasis in any conversation should be on the highest caliber movie that would be highly profitable for the studio.

Conclusion 5. Better presentations of engineers in the news media could also inform the public about infrastructure issues.

Several respondents felt that the news media could make better use of engineers as resources, such as quotable sources of information about the engineering aspects of a story. One respondent pointed out that every month major newspapers run many stories in the metro section that would benefit from a reporter who understood engineering. Subjects include proposed transportation systems, infrastructure failures, new construction, etc.

Conclusion 6. Most engineers recognize the need for the engineering profession to speak and recruit with one voice and with consistent messages that will reach a wide audience.

The engineering community is divided into many specialty societies and trade associations, which tend to be strongly territorial, and bringing them together into a coordinated campaign could be difficult. At the same time, there was a consensus among respondents that the engineering profession should advocate and recruit as one profession with one message to encourage math studies and to boost applications to engineering schools.

A first step has been made by the American Association of Engineering Societies (AAES), which sponsored a print ad campaign in 2001 based on the "quality of life" theme; in April 2002 AAES began a radio advertising campaign featuring "voices of innovation," descriptions of inventions that have changed our lives and our society. The 2001 ad campaign was prepared on a $250,000 grant from the United Engineering Fund, which did not include funds for pre- and post-ad testing. As a result, AAES has not been able to demonstrate the results of the campaign through changes in attitude or behavior. AAES has applied for a National Science Foundation grant to measure the results of the radio campaign.

Conclusion 7. Almost no measurements of long-term outcomes have been done for outreach programs. The success of future programs will require

measurements and modifications as necessary to ensure that problems are being addressed effectively.

One of the surprising findings of this study is that only a minuscule number of outreach programs objectively measure and evaluate their effectiveness. Most of them measure effectiveness by the number of people who take part, anecdotal conversations that indicate satisfaction with the program, or increased requests for participation or materials. Although these data points measure enthusiasm for programs, they do not measure whether, or how effectively, programs are changing attitudes or behaviors, such as improving math literacy, increasing engineering applications, or understanding more about the engineering profession and the value of engineering to society.

Respondents suggested two reasons for the lack of meaningful measurement: (1) the cost is prohibitive for many organizations that must struggle to find funding for programming and have none left for measurement; and (2) the grassroots nature of most outreach activities (e.g., engineers participating in local and regional programs and classrooms) makes measurement very difficult.

Setting measurable objectives for future programs will be critical. This will involve setting benchmarks or baselines for various criteria at the outset of a program so that changes can be measured as the program evolves. Decades of well-intentioned, enthusiastic outreach activities at the grassroots level have made little headway in improving public attitudes. An effective coordinated effort will require measurement for two important reasons: (1) demonstrated results will establish credibility and ongoing support from the engineering community; and (2) ineffective programs can be modified to solve problems more effectively.

MESSAGES

"Messages" are statements that have been developed for repeated use over time. People try hard every day to get our attention, but we can only absorb so many messages and information in 24 hours. Americans are bombarded with hundreds of messages every day from family, colleagues, employers, government entities, corporations, nonprofit organizations, community organizations, and educational institutions. Those messages come in many forms—conversation, post-it notes, mail, posters, telephone, hand-held wireless devices, movies, TV dramas and sitcoms, newspapers and magazines, books, e-mail, web sites, advertising, billboards, and others. Successful communications generally have some common characteristics, and one of them is a well-defined set of messages.

Successful messages are ones people respond to in action or thought. In fact, we ignore the vast majority of messages to which we are exposed during the day. We hang up on telemarketers, flip past advertising in newspapers and magazines, change the radio and television channel, discard unsolicited postal and electronic mail without reading it, and choose not to click on advertising boxes at web sites.

The organizations that are most successful at getting us to pay attention to their messages do everything in their power to increase the likelihood that their messages stand out to the people they want to reach. Messages are carefully developed and tested on selected audiences to determine if they are effective with that audience. Messages are then delivered via many channels to increase the likelihood that members of the target audience will see or hear them. And they are delivered on a regular and consistent basis, repeated consistently and repeatedly so people are exposed to them enough to finally accept them, understand them, and be influenced by them. Extensive research on the target audience is ongoing to make sure the messages are reaching them.

Developing messages about engineering for the general public has been challenging because the public understands very little about engineers and what they do. In research with focus groups for the American Society of Civil Engineers (ASCE) and the American Council of Engineering Companies (ACEC) by Market Research Bureau, it was learned that the language engineers use to describe their work is not understood by the public. That finding alone could explain why the public is not knowledgeable about engineering. Even the word "engineer" was hard for people to define because of the wide variety of people who have been labeled "engineers": sanitation engineers, facilities engineers, domestic engineers, Novell-certified engineers, computer engineers, genetic engineers, and so on. Focus group members indicated that the word engineer had been used in so many ways that it had ceased to have real currency as meaning someone in a respected and valued profession.

The phrase "the built environment" had no meaning for people, who had to be cued as to its meaning. A discussion of how engineers contribute to the quality of life brought the wrong reactions. The list of people who contribute to quality of life included babysitters, plumbers, garbage collectors, postal carriers, newspaper deliverers, and many others. Engineers were on the list but probably not in the top 50. As one respondent pointed out, quality of life is a state of mind that is based more on levels of traffic congestion than the utility of the bridge or road on which you are stuck. That same respondent noted that engineers make significant contributions to standards of living, which may be what engineers mean when they talk about quality of life. To the focus groups, attributing quality of life to engineers seemed to be overstating the case. In addition, the important audience of K-12 students have trouble relating to this message because they have little control over their quality of life and are not likely to be interested in the societal contributions of engineering.

The question remains what credible and compelling messages the engineering community can deliver. Respondents to the NAE questionnaire said that consistent messaging across the engineering community would be one benefit of a coordinated campaign. Some of the respondents provided thoughtful messages that would probably resonate with a variety of target audiences. The consultants (from Market Research Bureau and McMahon Communications), who had brainstorming sessions on engineering messages in the past, agreed that the messages suggested by respondents to the questionnaire include some that could be very effective in promoting engineering as a career and raising awareness of engineering.

The messages collected from the questionnaires can be tested with appropriate audiences for their receptivity and credibility. It is important that messages be thoroughly tested because the selected messages will underpin any communications efforts. Once the testing has been done, the message set can be disseminated within the engineering community, along with objective data to support its adoption by the community for all outreach activities. The testing and adoption of a set of messages would have a direct and immediate benefit to the engineering community.

Respondents suggested all of the familiar messages about engineering, but also some that expressed traditional messages in a new way or presented an entirely new perspective. The most promising of these are listed below. From these, we believe a compelling set of messages could be developed and used to the advantage of the entire engineering community. Each message is listed only once but might be appropriate for more than one audience.

Messages for Students

- An engineering education is valuable as the basis for a variety of careers.
- Engineering offers challenges, excitement, opportunities, and satisfaction.
- Engineering is worthwhile, challenging, fun, and within reach.
- An engineering career provides flexibility that allows for family life choices and helping people.
- It's not as complicated as you think. Anyone can understand the principles of engineering.
- Engineering is a collection of diverse fields that need people with diverse talents, experiences, creativity, and entrepreneurial spirit.
- Math (and science and technology) literacy will open doors in your future.
- Math is the alphabet of science and engineering.
- The excitement of engineering is that engineers create "something that has not been" for the good of humanity.
- Engineering includes a variety of fields of study and occupations.
- Math is challenging. Competition is not limited to sports.
- Engineers are not what people expect.
- Engineering is a springboard to many career opportunities.

Messages for Parents, Teachers, and Guidance Counselors

- Students must take the tough courses if they want to enter a college of engineering.

- Engineering features low unemployment, and engineers are creative and human.
- If you like math, you'll love engineering.
- If you like solving problems, wait till you hear what engineers can do.
- Engineering offers a lifetime of interesting work.

Messages for the Engineering Community (for internal use by trade associations and professional societies and for recruiting at the college level)

- Engineering is a core competency necessary to solve the complex technical and environmental challenges facing our customers and stakeholders.
- Engineering is a lion's profession.
- Engineering provides innovative solutions to societal problems.
- Engineers make use of both old and new knowledge to solve practical problems.
- Engineering plays an influential role in the burgeoning fields of bioethics, national security, and others.
- Engineering is a profession for leaders.

Messages for Policy Makers and Opinion Leaders

- Engineers are responsible for the high standard of living in the United States.
- Engineering builds societies. Engineering advances economies.
- Engineering is integral to society's progress and a country's ability to produce wealth.
- The United States can't expect to continue to solve its problems by importing technical talent from developing countries.
- Engineers are technology-literate citizens.
- Global competitiveness demands that many of the best and brightest students enroll in engineering.
- If engineering enrollments continue to decline, our nation will face a severe competitive and economic crisis.

REFERENCE

Mathias-Riegel, B. 2001. Engineering that's elementary. Prism 10(7):34-36.

4

Recommendations

Based on the results of their research, the NAE's consultants formulated a set of draft recommendations, which they first reviewed with a small advisory group drawn from organizations with members on the CPAE. Based on this feedback, a revised set of recommendations was drafted for the CPAE to consider at its meeting in Washington, D.C., on April 16, 2002.

The CPAE heard NAE Executive Officer Lance Davis outline the genesis of the Public Awareness of Engineering project. The consultants presented their survey findings and described selected current outreach activities by the engineering community. Dr. Davis then presented the draft recommendations. CPAE Chairman Steve Bechtel led the Committee's discussion of the recommendations and next steps.

The preliminary report was finalized and sent back to CPAE members for approval. The recommendations and next steps, as approved by the CPAE on June 24, 2002, are as follows:

SETTING THE STAGE

Public awareness is a driving force, not a guarantee of desired outcomes. Awareness is a necessary *milestone* towards desired outcomes. The inventory demonstrates that the engineering community is engaged in trying to improve public awareness. The engineering community recognizes that awareness currently is marginal and coordinated efforts would significantly improve effectiveness. It appears timely to capitalize on the collective desire to improve awareness.

GOAL: IMPROVED PUBLIC AWARENESS OF ENGINEERING

Improving public awareness of engineering will have the following outputs: more technical literacy among decision makers; more technical literacy in the general public; and more and better prepared students in engineering.

Improving awareness will lead to the following long-term outcomes: increased global competitiveness; improved public policy; increased national security; a public more intelligently engaged in technology issues that affect their lives; and an improved standard of living.

Achieving the Goal

There are two equally important methods to achieve the goal: In the short term, focus on public relations/public affairs. In the long term, focus on education.

Public Relations/Public Affairs

Objectives:

- Increase the understanding among target audiences of what engineers do and the role they play in our society. Target audiences include students, parents, teachers, guidance counselors, the media, policy makers, and the informed public.
- Increase the number of people who can "play back" a positive message about engineers when asked.
- Increase the use of engineers as information resources by journalists (who influence public opinion).
- Increase the use of engineers among decision makers and the informed public as resources on technology issues.
- Increase engineers' involvement in public policy decision making.

Recommendations: Public Relations/Public Affairs

- Current activities: Maintain the current interest level and participation in local grassroots efforts by the engineering community.
- Current Outreach Programs: Share with the engineering community.
- Consistent Messages: Determine effective messages (through testing) and encourage the engineering community to use them. Provide guidance nationally on effective use of messages.
- Develop nationally-coordinated grassroots efforts to reach new/larger audiences.
- Measure effectiveness of efforts: Encourage the engineering community to measure effectiveness of its initiatives targeting key audiences.

Initiatives for Consideration:

(Not prioritized and all should include developing measures of effectiveness)

- *Internet Information*
  Create a web-based "clearinghouse" for information about engineering outreach accessible not just to the engineering community, but to teachers, parents, students, media, etc.

- *Media Education Programs*
  These would include interaction with reporters at high-level fora, one-on-one meetings with influential media, and small group presentations for regional newspapers.

- *Movie and TV Show Development*
  These should be geared toward children.

- *Advertising and Public Service Announcements*
  A targeted ad campaign could be developed, but needs to be cost effective and should be unified through consistent messages, taglines, and logos. The campaign should generate awareness through focus on basic social issues: standard of living, global competitiveness, economic growth, and national security.

- *Public Lecture Series and Exhibits*
  Encourage lecture series and exhibits on engineering topics (perhaps in conjunction with engineering schools) on topics ranging from "hot" practice areas to new engineering developments to large projects in local areas, directed to a generally educated audience who are not engineers, and using engaging speakers.

- *Grassroots Outreach*
  Harness and build on the interest and enthusiasm of current outreach activities across the engineering community. Support these efforts by developing common messages, sharing experiences through a web-based clearinghouse, and providing guidance on objectives, execution, and measurement of effectiveness. Unify the efforts through consistent messages and increase efforts to bring activities to more schools and communities on a continuous basis throughout the year.

- *Competitions*
  Appeal to the competitive spirit of young people to generate interest in engineering.

Use existing local, regional and national competitions to generate more visibility and create more opportunities for greater interaction between students and engineers. Work to expand participation within the engineering community including industry, societies, national labs, and academia.

Education Intervention

Objective:

Strengthen the U.S. K-12 education system to include greater emphasis on engineering and technology, as well as math and science: **An *education* solution.**

Specific Objectives:

- Increase the numbers of high school seniors applying to engineering colleges.
- Increase the quality of applicants to engineering colleges.
- Increase the quality of math instruction at the K-12 level.
- Increase the quality of math comprehension at the K-12 level.
- Increase the awareness of engineering as a career option among elementary and middle school students.
- Increase the numbers of states that set standards of learning for engineering along with math and science.

Recommendations: Education Intervention

- Share information on existing local programs with the engineering community.
- Convene a summit of deans of engineering schools, education schools, and teacher groups to share experiences and brainstorm solutions on training the next generation of teachers.
- Commission a study to determine the dynamics of how children learn technical concepts and the gaps in the research addressing this area.
- Provide guidance on developing new K-12 curricula in engineering/technology/math/science.
- Develop a public policy program to influence decision makers on all levels—national, state, and local—to address education issues.

NEXT STEPS:

Next Steps: Public Relations/Public Affairs

- The entire engineering community, to include professional societies, industry, academia, government, and national laboratories, should come together to establish specific programs and mechanisms that will achieve the recommendations and objectives set forth above.
- NAE should convene an initial symposium with working groups to begin this process.

Next Steps: Education Intervention

- The engineering community should bring together a blue-ribbon council or conference of engineering, education, and public policy communities to develop a specific action plan and intervention strategies for changing the K-12 education system.
- NAE should begin this process by convening this blue-ribbon council.

Appendixes

Appendix A

Engineering Enrollments

As Figure A-1 shows, the number of B.S. degrees in engineering has waxed and waned over the last 55 years. The number peaked at 78,172 in 1986 and then declined through the 1990s to about 63,000. In 2001, the number rebounded slightly to 65,113. Engineering enrollments are trending up at the moment, so graduations may increase further if drop-outs do not exceed historical rates. The ratio of engineering B.S. degrees to the total number of bachelor's degrees has generally declined from about 10 percent in the 1950s to 6 percent in the 1990s (Figure A-2).

Decreasing enrollments through the 1980s and early 1990s caused great concern in the engineering community because they suggest a decreasing interest among young students and their mentors in engineering as a career choice. The concern was and is exacerbated by the lack of information about why enrollment declined or what it ought to be. The *National Science Foundation (NSF) 2000 Science and Engineering Indicators* suggest that the issue is largely demographics in that the college-age population as a whole declined from 21.6 million in 1980 to 17.0 million in 2000. However, the number of students enrolled in college peaked in 1992 at 14.7 million (and has been flat since), 10 years after engineering enrollments began to decline (NSB, 2000).

The H1b visa program has probably softened the market for engineering graduates somewhat in the 1990s, but the program began in 1990, eight years after the initial decline in engineering enrollments. Anecdotally, it has been suggested that some engineering schools began to emphasize Ph.D. programs during the 1980s, and, because they had fixed resources, including limited faculty, they compensated by limiting undergraduate enrollment. Another possible explanation is the response of engineering schools to the perceived strength or weakness

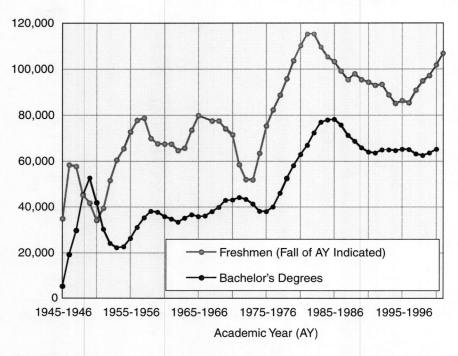

FIGURE A-1 Trends in College Enrollments and Graduations. Source: Engineering Trends, 2002.

in the job market in their areas. Even though the U.S. economy was very strong throughout the 1990s, global outsourcing of engineering was greatly increased. The impact this had on local job markets and a decision by some schools to restrict enrollments is not clear. Finally, the period in question coincides with the Information Technology revolution, which made engineering a more efficient activity.

Estimating future needs is, of course, even more problematic. To a first approximation, we can assume that all engineers with degrees who are 55 or older will retire in the next 10 years. According to the *NSF 2002 Indicators*, there are about 340,500 of them, but only about two-thirds of them (about 227,000) are in the science and engineering (S&E) workforce. The *2002 Indicators* projects that there will be 138,000 new engineering jobs between 2000 and 2010 (NSB, 2002). Interestingly, that is a smaller percentage increase than expected for the economy overall—10 percent versus 15 percent. In any event, 365,000 engineers will be needed by 2010. If present graduation trends continue, and if only two-thirds remain in the S&E workforce, about 420,000 engineers will be available. The demand for new graduates could decline further depending on the number of H1b visas issued and the amount of outsourcing. The numbers are very uncertain,

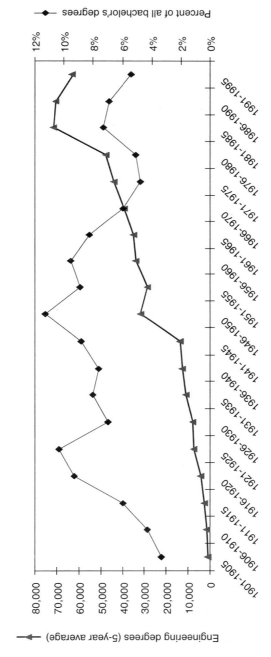

FIGURE A-2 Engineering Bachelor's Degrees Five-Year Average. Source: Adapted from Burton and Parker, Forthcoming.

however. The projected increase in engineering jobs could prove to be a gross underestimate.

In any case, the number of engineering graduates in the United States does not compare favorably with the number graduated by our major global competitors (see Table A-1). China produces three times as many as the United States, the European Union about twice as many, and Japan about two-thirds again as many. In terms of percentages, the ratio of engineering degrees to total undergraduate degrees in China is nearly nine times that of the United States, Japan four times, and the European Union three times. Thus, a substantially lower proportion of U.S. undergraduate students are studying engineering than the pro-

TABLE A-1 International Comparison of Engineering Degrees

		Total Degrees	Engineering Degrees	24-year-olds[a]	Engineering Degrees	
					Percentage of Total Undergraduates	Percentage of 24-year-olds[a]
US	1998	1,199,579	60,914	3,403,039	5.08%	1.79%
	1996	1,179,815	63,114	3,671,000	5.35%	1.72%
	1991[b]	1,107,997	62,187	3,584,000	5.61%	1.74%
Japan	1999	532,436	103,440	1,771,600	19.43%	5.84%
	1997	524,512	102,951	1,870,700	19.63%	5.50%
	1992[b]	400,103	81,355	1,787,400	20.33%	4.55%
China	1999	440,935	195,354	20,047,600	44.30%	0.97%
	1996	325,484	148,844	23,220,000	45.73%	0.64%
	1992[b]	308,930	112,814	25,428,000	36.52%	0.44%
EU	1999 - see note	1,908,967	134,692	4,903,035	7.06%	2.75%
	1997 - see note	1,070,238	139,020	4,975,100	12.99%	2.79%
	1992 - see note	604,551	95,594	5,548,880	15.81%	1.72%

Source: NSB, 1993, 2000, 2002.

[a]Data for 1991 and 1992 are for 22-year-olds.
[b]Data for 1992 data do not include Austria, Finland, and Sweden, which joined the European Union in 1995.

Note: NSB 1999 data for Austria, Belgium, Denmark, Finland, France, Germany, and the United Kingdom are from 1999; for Ireland, Italy, the Netherlands, Spain, and Sweden from 1998; for Portugal from 1996; and for Greece from 1993. NSB 1997 data are from 1997 for Austria, Denmark, Germany, the Netherlands, Sweden; for the United Kingdom from 1997; for Belgium, Finland, France, Ireland, Italy, Portugal, and Spain from 1996; and for Greece from 1993. NSB 1992 data from Austria, Finland, Greece, Sweden, the United Kingdom are from 1991; for Portugal from 1989; and for Belgium from 1988.

portion in our major international competitors. The United States produces the second fewest engineering degrees per year for its 24-year-old population. In this comparison, only China produces a lower fraction, but it still produces about three times as many engineers on an absolute basis.

U.S. students enrolling in engineering are overwhelmingly white males (Figure A-3). African Americans and Hispanics, who account for about 25 percent of the population, account for only about 6 percent of the engineering workforce and about 11 percent of engineering B.S. degrees. Women are also underrepresented. They account for only 19 percent of B.S. engineering graduates, although the number of men and women in many other degree programs is about equal. Only about 2 percent of female B.S. recipients graduate in engineering (Figure A-4). Nevertheless, while total enrollments have been basically flat, the number of African American, Hispanic, and women enrollments have risen steadily over the last 20 years. Thus, underrepresented minority students and women have replaced white males, rather than increasing the pool.

The preparation (or lack of preparation) of elementary and high school students for enrollment in engineering programs is an obvious source of concern. Some of the decline noted in Figure A-1, above, could be attributed to students recognizing that they do not have the necessary math and science background to succeed in the engineering curriculum. An indication of the alarming decline in performance among elementary and secondary students is shown in the Third International Math and Science Survey (NSB, 2002). In a comparison of performance with students in other countries, U.S. fourth graders had average scores in

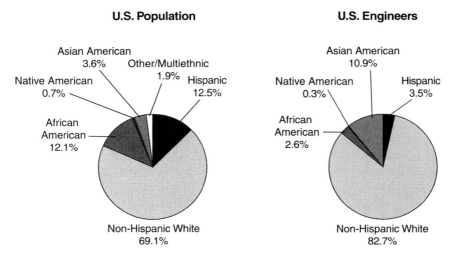

FIGURE A-3 Minorities in Engineering. Source: Adapted from U.S. Census Bureau, 2000; NSB, 2000.

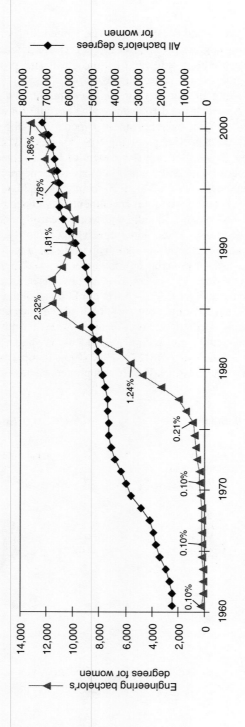

FIGURE A-4 Engineering Degrees Awarded to Women. Percentages indicate percentage of engineering degrees compared to total B.S. degrees earned by women in the U.S. Source: Adapted from Engineering Workforce Commission, 2000.

APPENDIX A

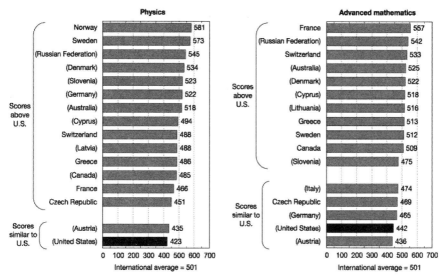

FIGURE A-5 Average scores on physics and advanced mathematics assessment for students in the final year of secondary school: 1994-1995. Source: Mullis et al., 1998.

math and well-above average scores in science. Eighth graders were 22 points below average in math and 9 points below average in science. Figure A-5 shows, by the twelfth grade, U.S. students are 60 to 70 points below average compared to students in many other countries.

REFERENCES

Burton, L., and L. Parker. Forthcoming. The Education and Employment of Engineering Graduates. Arlington, Va: National Science Foundation.

Engineering Trends. 2002. First-year Enrollments and BS Degrees. Available online at www.engtrends.com.

Engineering Workforce Commission of the American Association of Engineering Societies. 2000. Bachelor's degrees rising. Engineers 6(4):1-4.

Mullis, I., M. Martin, A. Beaton, E. Gonzalez, D. Kelly, and T. Smith. 1998. Mathematics and Science Achievement in the Final Year of Secondary School: ITEA's Third International Mathematics Study (TIMSS). Chestnut Hill, Mass.: Boston College, TIMSS International Study Center.

NSB (National Science Board). 1993. Science and Engineering Indicators–1993. NSB-93-1. Arlington, Va: National Science Foundation.

NSB. 2000. Science and Engineering Indicators–2000. NSB-00-1. Arlington, Va: National Science Foundation.

NSB. 2002. Science and Engineering Indicators–2002. NSB-02-1. Arlington, Va: National Science Foundation.

U.S. Census Bureau. 2000. Population and Housing Tables. Available online at www.census.gov/population/www/cen2000/phc-t1.html.

Appendix B

Sample Mission Statements

ENGINEERING SOCIETIES

- To contribute effectively in the shaping of public policy and public awareness of engineers and the engineering profession in the United States.
- To provide leadership and support for the national effort to increase the representation of successful African American, American Indian and Latino women and men in engineering and technology, math- and science-based careers.

COLLEGE/UNIVERSITIES

- Research, teaching, economic development, outreach, and service.
- Education, research, and societal service. Educate the engineers of the future to create capacity for our nation. Education, service to the community, economic development.
- The mission is to provide . . . an effective engineering outreach program in technology transfer and continuing education for the state and the nation.
- To serve society by extending and transmitting the accumulated knowledge in the fields of engineering, computing, and nursing and by providing an environment for effective teaching, learning, and critical thinking.

INDUSTRIAL COMPANIES

- To provide quality products and services to our customers to provide them with a competitive advantage. Our reward for doing this successfully is growth and profits.
- To provide quality products for our customers around the world.
- Maximize value (i.e., stock price) for existing shareholders.

DESIGN/CONTRACTING FIRMS

- To engage in and carry on a general engineering, contracting, and construction-management business and to do engineering, contracting, and construction-management work of every nature and description.

MUSEUMS

- To inspire people to discover and enjoy science using programs and exhibits that emphasize education through interaction.
- We seek to inspire life-long learning by furthering the public's understanding of and enjoyment of science and technology.
- To promote science literacy, life-long learning, and an appreciation of the sciences by providing innovative, educational, and recreational experiences.

MEDIA/PRODUCERS

- To be a leading provider of educational, informational, and entertaining products and services using all media.

APPENDIX C

Selected Outreach Programs

The survey/questionnaire revealed a wide range of outreach activities throughout the engineering community. These activities take many forms but can mostly be characterized as local cooperation and support for national programs. All of the current outreach activities underway are important, and the sponsoring organizations obviously have invested a great deal of time, interest, energy, and money in them. In some cases, these organizations provide financial support that makes programs possible; in other cases, their staff members volunteer to work with young people on projects or visit their classrooms. The level of outreach and the commitment to young people and local communities are impressive.

Some current programs are especially noteworthy because they involve interesting or unusual collaborations or because they represent innovative practices not apparent in most other programs. These programs are described below.

TV Production. WGBH public television in Boston has two projects on the drawing board: (1) the development of engineering segments for its daily program "Zoom"; and (2) the design of a new children's show focused on engineering called "E Games." WGBH also aired "Building Big," an educational series on engineering achievement produced by Larry Klein for WGBH. American Society of Civil Engineers (ASCE) worked closely with WGBH on "Building Big" and "Zoom." Klein, president of Production Group, Inc., also produced a program on the collapse of the World Trade Center buildings based on ASCE and Federal Emergency Management Agency research, which aired on "Nova" in April 2002.

In network television, FIRST Robotics is working with Disney on a made-for-TV movie starring Noah Wyle of "E.R." as a teacher who supervises a FIRST team at an inner-city school in California.

Outreach to Women and Minority Students. With funding from an NSF grant, the Miami Museum of Science collaborated with the Miami-Dade County Public School System and SECME, the Southeastern Consortium for Minorities in Engineering, a national organization established in 1975 by the engineering deans of seven southeastern universities, working to increase the pool of minority students prepared to enter and complete college programs in math, engineering, and science. Girls in 52 middle schools were provided access to resources, which included teacher training and parent involvement.

"Math Is Power" is a humorous, sophisticated print and radio public service campaign that conveys the message to American students, particularly students from minority communities, that they can and should pursue quality, college-track math and science classes in middle and high school and that parents should demand these classes for their children because the skills will prepare them for a variety of career options. The campaign is sponsored by the National Action Council for Minorities in Engineering.

The focus of a program by Adobe Corporation is basic education in underserved communities. Adobe has established long-term relationships with local schools; the company provides funds for college scholarships and teacher training, donates software, and provides tutors. An annual program brings high school students to Adobe's offices where they learn from employees how to interview for a job and what skills are required for employment. An "Invitation-Only" event is an annual reception for women and minority students in local colleges. Adobe's outreach programs convey consistent messages: stay in school, go to college, and consider pursuing a career in technology.

Nine universities in and around Detroit have formed a partnership with the metropolitan school system to bring hundreds of middle and high school students, most of them minority students, to campuses on a year-round basis to be coached in the math, science, and computer skills they will need to be accepted and succeed in college. The Detroit area business community provides funding to pay the college faculty members for their teaching time. The University of Michigan, Dearborn, is one of the engineering schools that participates in this Detroit Area Public Schools Advanced Placement (DAPSAP) Program.

EXCITE (Exploring Interests in Technology and Engineering), sponsored by IBM, is a summer camp for sixth and seventh grade girls with 20 locations around the world. The goal of the program is to encourage interest in math and science. The "3M Science, Training Encouragement Program (STEP)" for 30 years has brought minority high school students to 3M for science classes.

Women in Engineering Program (WEP) and Multicultural Engineering Program (MEP) are two formal programs at the University of Colorado, Boulder, that focus on recruiting and retaining women and minority students. The programs have set numerical goals for measuring success: their goal is to graduate 30 percent women and 12 percent underrepresented minority students of color by 2007.

The only women's school with an engineering department, Smith College, which has a rigorous academic program, focuses on the social purpose of engineering.

One hundred Intel Computer Clubhouses in underserved communities around the world will enable youths ages 8 to 18 to develop computer fluency. The program is being established in partnership with the Boston Museum of Science and the MIT Media Lab.

Women in Engineering Programs and Advocates Network (WEPAN) has provided training to 175 universities focused on outreach, recruitment, and retention of women in engineering. WEPAN developed the curriculum and has conducted 2-1/2 day seminars over the past nine years to train universities to implement and evaluate recruitment and retention programs.

Advertising. ExxonMobil writes and places advertorials in *The New York Times* and *The Washington Post* to reach opinion leaders on issues related to science and technology. An advertorial entitled "In Praise of Engineering," which ran last year around E-Week, received a great deal of positive feedback from the engineering community.

Teacher Training. The Boston Museum of Science sponsors training for teachers in math and science. The Exploratorium Museum's Teacher Institute in San Francisco works with new middle and high school math, science, and physics teachers to increase retention rates.

Teach to the Future, a three-year program sponsored by Intel in partnership with Microsoft, will train 400,000 teachers in 20 countries to use technology and professional software in the classroom. 3M has a 25-year-old project, "3M Teachers Working in Science and Technology (TWIST)," which provides six-week summer internships for math and science teachers.

The engineering school at the University of Massachusetts, Lowell, offers a graduate-level class called "Engineering for Teachers" for middle and high school math, science, and technology teachers to teach them about the engineering profession and to give them the tools to introduce hands-on engineering design concepts into their curricula. The engineering and education schools at Tufts are working to improve training for math teachers in a program they hope will be a model for other universities.

Schools and Coursework. The Henry Ford Academy, a charter school developed by the Ford Motor Company with the Wayne County public school system, is located on the grounds of the Henry Ford Museum in Dearborn, Michigan. Entrance is by lottery, and the diverse student body, 70 percent minority, has needs ranging from special education to remedial learning to exceptional skills. Another program, sponsored by The Ford Academy of Manufacturing Science, has developed a pre-college curriculum used in 70 public schools nationwide, as well as in South Africa and India. The curriculum uses real-world settings for applied math and science instruction and employment-oriented courses on teamwork and communications. Courses are taught as electives along with the core offerings of each school system.

The GM Technical Academy was created by General Motors with the Oakland County Public Schools. During the school year, 20 high school juniors and seniors spend each morning at their own schools and each afternoon at GM; each year the students design, engineer, and build a full-size electrical vehicle (parts are fabricated by GM to the students' blueprints). Students, who range from degree-bound students interested in engineering to non-degree-bound students who want to pursue automobile-related vocational training, are also offered six-week paid internships at GM.

Space Day is an annual event that promotes math, science, and technology education by posing space-based problems for fourth, fifth, and sixth grade students to solve. This Lockheed-Martin program, which involves NASA, challenges students to work together in the classroom to solve design challenges for living and working in space. The day-long program also features a live, interactive Webcast (via the Internet or satellite) that allows students to interview astronauts, scientists, and space experts.

"Preparing Future Workforce" is a collaborative effort between 3M, the Saint Paul public schools, and the Saint Paul Area Chamber of Commerce that brings the concept of career pathways and how to pursue them into area high schools.

The Summer Program for Engineering, Math, and Computer Science, co-sponsored by the University of Vermont (UVM) and the Governor's Institutes of Vermont, brings high school sophomores and juniors from all over the country to campus. About 10 percent of the students who participate enter UVM as freshmen.

Mentoring. The ACE Mentor Program encourages students to pursue careers in architecture, construction, and engineering. Founded by Thornton-Tomasetti Engineering, the program, which is supported by sponsors, including ASCE, relies on more than a dozen large firms to provide mentors to work with students.

MentorNet, a Silicon Valley-based nonprofit program, is an email and Internet network that links women students in engineering and science with professionals in these fields.

Competitions. The annual National Science Bowl, a math and science competition among teams of high school students from around the country, is sponsored by the U.S. Department of Energy. Teams compete at the regional level, and the finalists come to Washington, D.C., for the national competition. The Idaho National Engineering and Environmental Laboratory (INEEL) works with teams in the state of Idaho through the INEEL Scholastic Tournament which tries to get every high school in Idaho to field a team. Every year, 80 to 85 teams are fielded in Idaho, more than in any other state; state championship matches are televised on Idaho public television. Three Idaho teams then advance to the national competition in Washington, D.C., giving students a chance to meet federal scientists and mathematicians.

MATHCOUNTS, a program sponsored by the National Society of Professional Engineers, focuses on honing math skills at the middle school level. The program offers a platform for organizations that go the extra step. The engineering school at the University of Vermont (UVM), for example, has made MATHCOUNTS part of its outreach program; in 2001 all four Vermont "mathletes" placed in the top 100 students for the first time. UVM also collaborates with the business community in Vermont in its annual Design TASC (Technology and Science Connection) Competition for high school students. The competition is sponsored by the Ford Motor Company and businesses in Vermont and surrounding states, including local TV and radio stations, which broadcast commercials and announcements at no cost to UVM.

Public Service Campaigns. The National Association of Manufacturers joined with the U.S. Department of Commerce in 2001 to co-sponsor a program to address the shortage of manufacturing workers in the United States; the program is funded by $1 million in in-kind donations. Information packets focusing on math and science education will be mailed to every middle school in the country. Televised Public Service Announcements, which will feature Mia Hamm, the Backstreet Boys, and a NASCAR driver, will focus on the manufactured equipment that enables them to play soccer, play music, and drive race cars.

Public Lecture Series. Harvey Mudd College sponsors the Dr. Bruce J. Nelson '74 Distinguished Speaker Series, which bring noted speakers to campus to inform the general public about scientific subjects; the series is popular with the public and press.

Public Policy Development. Kansas State University (KSU) has instituted a program to inform public policy makers about the contributions of engineers to society and the costs to the university of providing engineering education and conducting research. Its message to state legislators focuses on the contributions of engineers to the standard of living and economy of Kansas, and the significant contributions of KSU to the state's well-being. The object of the program is to influence state legislators to consider these contributions in funding decisions.

Education Standards. Massachusetts is the first state to have revised the K-12 math standards as math and engineering standards, introducing engineering and problem-solving concepts into the classroom in all grades. Many other states are watching closely with an eye to following suit. The engineering school at Tufts University has worked closely with the Massachusetts Department of Education, the Boston Museum of Science, and other institutions on this initiative. An NSF grant will provide funding for training current teachers to teach to the new standards.

Classroom and Teacher Support. An NSF grant enabled North Carolina State University to send nine engineering students to work with two elementary schools and one middle school as science, math, and technology resources and co-teachers. In Massachusetts, the Engineering in Mass Collaborative, a partnership of businesses, colleges and universities, K-12 teachers and schools, and state agencies, seeks out and promotes best practices in increasing awareness of engineering and science careers and improving math, engineering, science, and technology education for K-16 students. The engineering school at University of Massachusetts, Lowell, which founded the collaborative, directs the program.

National Engineers Week, sponsored by the engineering societies and industry on a rotating basis, brings engineers into classrooms at all levels, introducing many children to the world of engineering for the first time.

The Jason Program, founded by Bob Ballard, the scientist who located the Titanic, organizes an annual expedition and enables students in centralized locations in the United States and Canada to interact with the explorers. The Idaho National Engineering and Environmental Laboratory (INEEL) has taken on a leadership role in the state on this program, enabling students to travel to the centralized locations and facilitating the link-up, thus providing access for students in this largely rural state.

Saturday Morning Physics at Fermi National Accelerator Laboratory, which began in 1980, brings 300 high school students to Fermi 10 times a year. Students attend lectures on physics topics and then break up into discussion groups with physicists and post-doctoral researchers. QuarkNet, also at Fermi, is a multi-year teacher outreach program that has reached 720 high school physics teachers to date. The program brings teachers to the laboratory for summer research appointments; the teachers then provide professional development for other teachers through Web-based follow-on programs.

Grassroots Outreach Programs. The largest number of respondents engaged in outreach activities provide local, hands-on support for national programs. The national program provides the overall structure and relies on the participating organizations to provide expertise, enthusiasm, and time. National hands-on programs that were mentioned by respondents are listed below:

ACE Mentor Program
AICHE (American Institute of Chemical Engineers) kits for ages 8 to 10 (designed to help Girl Scouts earn two engineering badges, but also appropriate for classroom instruction)
ASCE/West Point Online Bridge Design

Boy Scouts of America (Engineer Explorer posts, open to girls and boys, co-sponsored by the Learning for Life Program)
Girl Scouts USA (relationships with AICHE, Society for Women Engineers)
SAE (Society of Automotive Engineers) Collegiate Design Competitions
Formula SAE
SAE MiniBaja
SAE Aero Design
SAE Clean Snowmobile
SAE's A World in Motion (grades 4 to 6)
International Bridge Competition
Jason Foundation for Education
FIRST Lego League/FIRST Junior Robotics
US FIRST Robotics
National Engineers Week

APPENDIX D

Committee on Public Awareness of Engineering

Stephen D. Bechtel, Jr. (Committee Chair), Chairman Emeritus and Director, Bechtel Group, Inc.
Norman R. Augustine, Chairman and CEO (retired), Lockheed Martin Corporation
William F. Ballhaus, Jr., President and CEO, The Aerospace Corporation
Craig R. Barrett, President and CEO, Intel Corporation
David L. Belden, Executive Director, American Society of Mechanical Engineers
Sherwood L. Boehlert - R-NY, U.S. House of Representatives
G. Wayne Clough, President, Georgia Institute of Technology
Vance D. Coffman, Chairman and CEO, Lockheed Martin Corporation
James E. Davis, Executive Director, American Society of Civil Engineers
Nicholas M. Donofrio, Senior Vice President, Technology and Manufacturing, IBM
E. Linn Draper, Jr., Chairman of the Board, President and CEO, American Electric Power
Thomas E. Everhart, President Emeritus, California Institute of Technology
Daniel S. Goldin, Former Administrator, NASA, Senior Fellow, Council on Competitiveness
Irwin M. Jacobs, Chairman and CEO, Qualcomm, Inc.
Ruben F. Mettler, Retired Chairman and CEO, TRW. Inc.
Judith Ramaley, Assistant Director of Education and Human Resources, National Science Foundation
Lee R. Raymond, Chairman and CEO, Exxon Mobil Corporation
Daniel Senese, Executive Director, IEEE
John Brooks Slaughter, President and CEO, National Action Council for Minorities in Engineering

David Swain, Senior Vice President, Boeing Corporation
Charles M. Vest, President, Massachusetts Institute of Technology
Wm. A. Wulf, President, National Academy of Engineering
William D. Young, Chairman and CEO, ViroLogic, Inc.

ATTENDEES AT CPAE MEETING

Alice P. Gast, Vice President for Research and Associate Provost, Massachusetts Institute of Technology
Attending on behalf of C. M. Vest

David Goldston, Chief of Staff, House Science Committee
Attending on behalf of S. L. Boehlert

W. R. K. Innes, President, ExxonMobil Research and Engineering
Attending on behalf of L. R. Raymond

Kevin Kelley, Senior Vice President, External Affairs, QUALCOMM Incorporated
Attending on behalf of I. M. Jacobs

Robert Spitzer, Vice President, Technical Affiliations & Univ. Relations, Boeing Company
Attending on behalf of D. Swain

David Tennenhouse, Vice President, Corp. Technology Group and Director, Research, Intel Corporation
Attending on behalf of C. R. Barrett

STAFF

Lance Davis, Executive Officer, National Academy of Engineering
Robin Gibbin, Director, Public Understanding of Engineering, National Academy of Engineering
Maribeth Keitz, Senior Information Assistant, National Academy of Engineering

ADVISORY GROUP

Lance Davis, Chair, Advisory Group, Executive Officer, National Academy of Engineering
Randy Atkins, Senior Media Relations Officer, National Academy of Engineering
Dennis Boxx, Vice President of Communications, Lockheed Martin Corporation

Casey Dinges, Managing Director, Communications, American Society of Civil Engineers
Doug Harrison, Manager Process Engineering, Exxon Mobil Corporation
Jane Howell, Director of Communications, American Society of Civil Engineers
Maria Ivancin, Principal, Market Research Bureau, LLC
Kirk D. Kolenbrander, Special Assistant to the President and Chancellor, Massachusetts Institute of Technology
Paige McMahon, Principal, McMahon Communications
Bernie Meyers, Senior Vice President, Bechtel Group, Inc.
Pat Natale, Executive Director, National Society of Professional Engineers
Russ Pimmel, Program Officer, Division of Undergraduate Education, National Science Foundation
Tom Price, Executive Director, American Association of Engineering Societies
Paul Torgersen, John W. Hancock, Jr. Chair of Engineering and President Emeritus, Virginia Polytechnic Institute and State University, NAE Council Representative

Appendix E

Engineering Communications, Education, and Outreach Questionnaire

Thank you for participating in our study. We would like to assure you that all of your responses will be kept confidential and used only in the aggregate with the responses from other organizations. We are asking you to identify your organization and yourself only to keep track of who has responded and to contact you if we need any clarification or additional information.

Please respond to the questionnaire for your entire organization. If you are not the person who is best able to respond, please pass the questionnaire along to that person. If you feel that another part of your organization should respond in addition to you, please advise us at the number below and we will send them a separate questionnaire.

Although we have tried to simplify the questionnaire with response categories for many of the questions, we encourage you to provide any explanation or additional information for any of the questions. If you have any difficulty in completing the questionnaire, please contact Maria Ivancin at Market Research Bureau – telephone: 202-789-2110; email: mivancin@sprintmail.com. If you would like to download the questionnaire and submit it electronically, you may do so at **www.nae.edu/engineeringsurvey**.

Please return the questionnaire by **December 7, 2001.** Thank you for your time and your participation.

BACKGROUND ON YOUR ORGANIZATION

1. What is the full name of your organization?
   _____

APPENDIX E 75

2. What type of organization is it?

❑ Professional engineering society
❑ Engineering company
❑ Corporation/private industry
❑ Academic institution
❑ Educational/academic association
❑ Curriculum development organization
❑ Government agency
❑ Charitable foundation
❑ Trade association
❑ Media
❑ Other PLEASE DESCRIBE

_____

3. How would you describe your organization's mission?

_____

_____

_____

COMMUNICATIONS, EDUCATION, AND OUTREACH EFFORTS

We are defining **"communications, education, and outreach"** efforts as any activities that are designed to help the public understand the role of engineering in society. We are interested primarily in the more formal activities that your organization might have (i.e. planned, designed and funded activities) but would welcome your comments on any informal activities that you feel might be relevant.

Current Efforts

1. Is your organization *currently* engaged in communications, educational, or outreach activities designed to improve the public understanding of engineering?

❑ Yes
❑ No – SKIP TO QUESTION 6 BELOW

2. Are these ongoing activities or discrete programs with limited duration?

   ❏ Ongoing only – SKIP TO QUESTION 5
   ❏ Discrete programs only ——— CONTINUE
   ❏ Both ongoing and discrete —— CONTINUE

3. How long have these activities been underway?

   _____
   INDICATE NUMBER OF YEARS OR MONTHS

4. How much longer are these activities going to last?

   _____
   INDICATE NUMBER OF YEARS OR MONTHS

5. Does your organization have more than one program of such activities (i.e. that reaches different audiences, has different objectives or different messages)?

   ❏ Only one
   ❏ More than one – How many programs does your organization currently have? _____

PLEASE SKIP TO QUESTION 8

Past Activities

6. IF YOUR ORGANIZATION IS *CURRENTLY NOT* ENGAGED IN ANY COMMUNICATIONS OR OUTREACH ACTIVITIES: Has your organization done any such activities in the past, that you are aware of?

   ❏ Yes
   ❏ No – IF YOUR ORGANIZATION HAS *NEVER* ENGAGED IN ANY COMMUNICATIONS OR OUTREACH ACTIVITIES PLEASE SKIP TO QUESTION 10 BELOW

7. When was the last time that your organization did any communications, education, or outreach?

   ❏ Within the past year
   ❏ 1-2 years ago
   ❏ 2-3 years ago
   ❏ 3-4 years ago
   ❏ 4-5 years ago
   ❏ More than five years ago

8. IF YOUR ORGANIZATION'S CURRENT PROGRAMS HAVE ONLY BEEN IN PLACE LESS THAN TWO YEARS: Was the current communications or outreach effort your organization's first effort or have there been others?

   ❏ First effort
   ❏ Have had other efforts

9. For how long has your organization had some sort of communications, education, or outreach effort?

   ❏ For less than 3 years
   ❏ 3-5 years
   ❏ 5-7 years
   ❏ 7-10 years
   ❏ For more than 10 years

FUTURE ACTIVITIES

10. Does your organization have plans for any communications, education or outreach activities in the future (beyond what you might be doing right now)? If you are doing something currently, do you have any plans for <u>other</u> efforts once the current activities are completed?

    ❏ Have plans for future (additional) communications, education, or outreach activities
    ❏ Do not have any plans for such (additional) activities

11. What would be the earliest date that such activity would begin?
    PROVIDE DATE _____

    IF YOU HAVE *NEVER* ENGAGED IN ANY COMMUNICATIONS, EDUCATION, OR OUTREACH PROGRAMS AND *HAVE NO PLANS* TO DO SO IN THE FUTURE PLEASE SKIP TO THE SECTION TITLED **"OTHER COMMUNICATIONS PROGRAMS"** ON PAGE 10.

Objectives

12. Please describe the **objectives** for the communications, education, or outreach efforts that you are engaged in. What are the reasons for engaging in these efforts? IF YOU HAVE MORE THAN ONE PROGRAM, PLEASE ANSWER SEPARATELY FOR EACH IF THE OBJECTIVES ARE DIFFERENT. PLEASE ATTACH ADDITIONAL PAGES AS NECESSARY.

    PROGRAM #1:
    _____
    _____

    PROGRAM #2:
    _____
    _____

    PROGRAM #3:
    _____
    _____

13. What do you expect as the outcome for these efforts? PLEASE ANSWER FOR EACH PROGRAM YOU HAVE IF YOU HAVE MORE THAN ONE. PLEASE ATTACH ADDITIONAL PAGES AS NECESSARY.

PROGRAM #1:

_____

_____

PROGRAM #2:

_____

_____

PROGRAM #3:

_____

_____

14. What are the target audiences for your organization's communications, education or outreach efforts? Who are you trying to reach with your messages? CHECK ALL THAT APPLY FOR EACH PROGRAM.

	PROGRAM #1	PROGRAM #2	PROGRAM #3
General public	❏	❏	❏
Engineers	❏	❏	❏
Potential clients/engineering users	❏	❏	❏
Undergraduate students	❏	❏	❏
Children – in Kindergarten through fifth grade (K-5)	❏	❏	❏
Children – in sixth through eighth grade (6-8)	❏	❏	❏
Children – in ninth through twelfth grades (9-12)	❏	❏	❏
Teachers in Kindergarten through twelfth grades (K-12)	❏	❏	❏
College/university faculty	❏	❏	❏
Opinion leaders	❏	❏	❏
Public policy makers	❏	❏	❏
Newspapers	❏	❏	❏
Broadcast media	❏	❏	❏
Other media PLEASE DESCRIBE _____	❏	❏	❏
Other audience PLEASE DESCRIBE _____	❏	❏	❏

15. What are the key messages you are trying to communicate through this communications or outreach effort? PLEASE ANSWER FOR EACH PROGRAM YOU HAVE IF YOU HAVE MORE THAN ONE. PLEASE ATTACH ADDITIONAL PAGES AS NECESSARY.

PROGRAM #1:
_____
_____

PROGRAM #2:
_____
_____

PROGRAM #3:
_____
_____

SPECIFIC ACTIVITIES

Please answer the following for the *current* program(s) that your organization is engaged in. If you do not currently have any active efforts, please answer for the *most recent* program.

We would also like to request samples of whatever activities you might be able to share with us, including videos, copies of ads, press kits, press releases, curriculum, direct mail pieces, etc. Please include what you can in the return envelope provided. If there are other pieces that you would like to share with us that will not fit in the envelope or that might not be covered by the postage, please check this box ❑ and we will make arrangements to get those from you.

1. Which of the following specific activities is your organization using in its communications or outreach efforts? CHECK ALL THAT APPLY

   ❏ Paid advertising
   ❏ Public service advertising (PSA)
   ❏ Direct mail
   ❏ Public relations (i.e. non-paid media activities)
   ❏ Public affairs/Public policy activities
   ❏ Web site(s)
   ❏ Speakers/symposia/forums
   ❏ 800 number
   ❏ Formal educational curriculum
   ❏ Informal educational programs
   ❏ Other PLEASE DESCRIBE: _____

   PLEASE COMPLETE THE APPROPRIATE SECTIONS BELOW FOR EACH TYPE OF ACTIVITY YOUR ORGANIZATION ENGAGES IN.

PAID OR PUBLIC SERVICE ADVERTISING:

2. What media do you use?

   ❏ Network broadcast television
   ❏ Broadcast local or spot television
   ❏ National cable television
   ❏ Local cable television
   ❏ Network radio
   ❏ Local or spot radio
   ❏ Outdoor/billboards/out-of-home
   ❏ Consumer magazines
   ❏ Trade or professional magazines/publications
   ❏ Newspapers
   ❏ Internet
   ❏ Other PLEASE DESCRIBE: _____

   PLEASE PROVIDE US WITH SAMPLES OF ANY MATERIALS.

3. FOR PAID ADVERTISING ONLY: Approximately what is your organization's advertising media budget?

_____
INDICATE MEDIA BUDGET

4. FOR PAID AND PSA: Approximately what is your budget for production?

_____
INDICATE PRODUCTION BUDGET

5. What call to action, if any, do you have in your advertising? What, if anything, do you ask the audience to do? (For example, call an 800 number or visit a web site for more information).

_____

_____

_____

PUBLIC RELATIONS:

6. Do you target any particular media?

❏ Television
❏ Radio
❏ Consumer magazines
❏ Trade or professional magazines/publications
❏ Newspapers
❏ Web-based media
❏ Other PLEASE DESCRIBE:
_____

7. Do you have a press kit?

❏ Yes – PLEASE PROVIDE US WITH A SAMPLE
❏ No

8. Have you scheduled any meeting(s) with a news organization(s) in the past year to explain the work of your organization or institution?

   ❏ Yes, have scheduled meetings with news organization(s)
   ❏ No, have not scheduled meetings with news organization(s)

PUBLIC AFFAIRS/PUBLIC POLICY:

9. What type of public policy or public affairs activities does your organization engage in?

   ❏ Lobbying
   ❏ Briefings
   ❏ Other PLEASE DESCRIBE:
   _____

10. Have you had meetings with elected officials to explain your work and/or resources?

    ❏ Yes, have had meetings with elected officials
    ❏ No, have not had meetings with elected officials

WEB SITE(S):

11. Is the web site your organization uses in its communications effort specifically designed for that effort or is the site also used for other reasons (e.g. member information)?

    ❏ Same web site
    ❏ Different web site

12. IF THE SAME: Is there a separate link or icon on the site directed at the target audience for your communications effort?

    ❏ Separate link/icon for target audience
    ❏ No link/icon for target audience

13. Is there a link or icon for the media or press on your organization's web site?

    ❏ Separate link/icon for media
    ❏ No link/icon for media

14. What is (are) your web address(es)?

   _____

   _____

   _____

   _____

SPEAKERS/SYMPOSIA/FORUMS:

15. Please indicate whether you participate in and/or sponsor each of the following. Also indicate how many of each type of activity your organization may have done in the last year, as well as the target audience(s) for each type of activity.

	Participate Only	Sponsor Only	Participate AND Sponsor	# in last year	Target Audience(s)
Speakers	❑	❑	❑	_____	_____
Symposia	❑	❑	❑	_____	_____
Forums	❑	❑	❑	_____	_____

16. Do you have speakers on staff for this purpose?

    ❑ Yes, have speakers on staff
    ❑ No speakers on staff

EDUCATIONAL PROGRAMS FOR ENGINEERING & TECHNOLOGY:

17. What grades are your programs designed for?

    ❑ K-5
    ❑ 6-8
    ❑ 9-12
    ❑ College
    ❑ Graduate school
    ❑ Adult

18. Which of the following activities do you provide?

    ❏ Formal curriculum materials
    ❏ Materials to guidance counselors
    ❏ Competitions
    ❏ Mentor programs
    ❏ School-to-work training materials
    ❏ National Engineers Week
    ❏ Other PLEASE DESCRIBE:
    _____

19. Did your organization develop the curriculum/materials or did you get it/them someplace else?

    ❏ Developed by your organization
    ❏ Developed elsewhere – From whom did you get the materials? Who were the materials developed by?
    _____

20. Are the educational materials you use specifically aimed at educating students about engineering and technology distinct from science?

    ❏ Yes, specifically designed to educate about engineering
    ❏ No, engineering is part of science materials

    PLEASE PROVIDE US WITH SAMPLES OF YOUR MATERIALS.

MONITORING AND EVALUATION

1. How do you determine that your communications or outreach efforts are successful? IF YOU HAVE MORE THAN ONE PROGRAM, PLEASE ANSWER SEPARATELY FOR EACH. PLEASE ATTACH ADDITIONAL PAGES AS NECESSARY.

PROGRAM #1:
_____

_____

PROGRAM #2:
_____

_____

PROGRAM #3:
_____

_____

2. What criteria are important in determining if your communications effort is successful?

	PROGRAM #1	PROGRAM #2	PROGRAM #3
Awareness	❏	❏	❏
Knowledge	❏	❏	❏
Attitudes	❏	❏	❏
Change in behavior	❏	❏	❏
Other PLEASE DESCRIBE _____	❏	❏	❏

3. Which, if any, of the following do you use in evaluating your programs?

❏ Tracking research (market research to track awareness or attitudes)
❏ Placement reports (for public service advertising)
❏ Response tracking (for direct mail or direct response advertising)
❏ News coverage in print, broadcast and/or online media
❏ Exit interviewing or on-site evaluations (for speakers, symposia, forums)
❏ Web site hits
❏ Other PLEASE DESCRIBE:
_____

4. Do you feel your efforts have been successful in terms of meeting the original objectives?

	PROGRAM #1	PROGRAM #2	PROGRAM #3
Successful	❏	❏	❏
Unsuccessful	❏	❏	❏

5. Why do you say that?

_____

_____

_____

6. IF YOU FEEL ANY OF YOUR PROGRAMS HAVE NOT BEEN SUCCESSFUL: Why do you feel they have fallen short of your objectives? What would have made them more effective?

_____

_____

_____

PLANNING PROCESS

These questions deal with how your communications, education, or outreach program is planned. Please be assured that your responses will be kept in strictest confidence.

1. Does your organization have a department dedicated to communication, education, or outreach?

   ❏ Have communications, education, or outreach department
   ❏ Do not have such department – SKIP TO QUESTION 4

2. How big is that department? How many people are in the department?

_____
INDICATE NUMBER OF STAFF IN DEPARTMENT

3. What is the full name of that department?

_____

4. Does your organization use any of the following in your communications or outreach efforts?

   ❑ Advertising agency
   ❑ Public relations firm
   ❑ Independent consultant in advertising/PR
   ❑ Educational consultant
   ❑ Other outside consultant PLEASE SPECIFY _____

5. Who within your organization is involved in decisions regarding things like objectives, messages and target audience for your communications or outreach efforts?

   ❑ Only you
   ❑ Communications/outreach department
   ❑ Finance/accounting department
   ❑ Your organization's management
   ❑ Board of directors
   ❑ Other PLEASE SPECIFY _____

6. What is the overall annual budget for your communications or outreach efforts?

_____
INDICATE ANNUAL BUDGET

7. Does this include internal staffing?

   ❑ Includes staffing
   ❑ Does not include staffing

APPENDIX E

8. Who determines this budget?

   ❏ Only you
   ❏ Communications/outreach department
   ❏ Finance/accounting department
   ❏ Your organization's management
   ❏ Board of directors
   ❏ Other PLEASE SPECIFY _____

9. How is the budget determined?

   ❏ Based on need
   ❏ Based on previous year
   ❏ Based on industry data
   ❏ Percentage of overall organizational budget
   ❏ Other PLEASE SPECIFY _____

10. Do you have a *written* public communications, education or outreach plan?

    ❏ Yes, have written communications plan – PLEASE SHARE THIS PLAN WITH US IF YOU CAN.
    ❏ No, do not have written plan

OTHER COMMUNICATIONS PROGRAMS

1. Are you aware of any other organizations that are engaging in communications, education, or outreach programs on engineering?

   ❏ Aware of other programs
   ❏ Not aware of other programs – SKIP TO QUESTION 5

2. Which other programs are you aware of?

   _____

   _____

3. Are there any programs that you are aware of that you feel are particularly effective in helping the public understand the role of engineering in society? Which ones do you feel are effective?

_____

_____

4. Why do you feel they are effective?

_____

_____

5. Are you partnering or coordinating with any other organization for any of your current or most recent communications or outreach efforts?

❑ Partnering/coordinating with other organization
❑ Not partnering – SKIP TO QUESTION 7

6. Which organization/s?
_____

7. Have you *ever* partnered or coordinated with any other organization in any of your communications efforts?

❑ Have partnered/coordinated with other organization
❑ Never partnered – SKIP TO NEXT SECTION "COORDINATED COMMUNICATIONS EFFORTS"

8. What organizations have you partnered with?

_____

_____

COORDINATED COMMUNICATIONS EFFORTS

1. Do you feel there is a need for coordinated communications efforts that would encourage consistent messages and possibly provide efficiency for various organizations interested in communicating about engineering?

   ❏ There is a need for a coordinated effort
   ❏ There is no need for such an effort

2. Why do you say that?

   _____

   _____

   _____

3. If there were a coordinated effort, what do you feel are the most important messages to communicate about engineering?

   _____

   _____

   _____

4. Would your organization be interested in participating in such coordinated communications efforts? (NOTE: This does not obligate you in any way. We are just interested in your opinion at this point.)

   ❏ Would be willing to participate
   ❏ Would not be willing to participate

5. Why do you say that?

   _____

   _____

   _____

ABOUT YOU

These last few questions will help us in our analysis of the results.

1. What is your name? _____

2. What is your title? _____

3. How long have you been employed in your current position?
   _____
   INDICATE # OF YEARS

4. How long have you been in your current industry?
   _____
   INDICATE # OF YEARS

5. Please provide your phone number and email in case we need to reach you.

   Telephone: _____

   Email: _____

Thank you for your time in completing this questionnaire. Please return the questionnaire in the envelope provided. Also please include any materials that you can share with us to help us better understand your communications activities. We have provided postage to cover the questionnaire and some limited additional materials. If you would like to share materials that do not fit into this envelope or may not be covered by the postage, please check this box ❏ and we will make arrangements to get those from you.

Once we have completed our study we will share our results with you. Thank you.

APPENDIX F

List of Organizations Responding to NAE Inventory Questionnaire

ENGINEERING SOCIETIES

AACE International
Accreditation Board for Engineering and Technology
Air and Waste Management Association
American Academy of Environmental Engineers
American Association of Engineering Societies
American Chemical Society
American Council of Engineering Companies
American Institute of Aeronautics and Astronautics
American Institute of Architects
American Institute of Chemical Engineers
American Institute of Mining, Metallurgy & Petroleum Engineers
American Physical Society
American Public Works Association
American Society for Engineering Education
American Society for Engineering Management
American Society of Agricultural Engineers
American Society of Certified Engineering Technicians
American Society of Civil Engineers
American Society of Heating, Refrigerating and Air Conditioning Engineers
American Water Works Association
ASM International
ASME International
Associated Social and Foundation Engineers

Association for Facilities Engineering
Board of Certified Safety Professionals
Civil Engineering Research Foundation
Commission on Professionals in Science and Technology
Construction Specifications Institute
Gateway to Educational Materials
Human Factors and Ergonomics Society
Institute of Electrical and Electronics Engineers, Inc.
Industrial Research Institute
Institute for Operations Research and the Management Sciences
Institute of Industrial Engineers
Institute of Transportation Engineers
Iron and Steel Society
ISA The Instrumentation, Systems, and Automation Society
Junior Engineering Technical Society (JETS)
National Academy of Building Inspection Engineers
National Academy of Engineering
National Academy of Forensic Engineers
National Action Council for Minorities in Engineering
National Council of Examiners for Engineering & Surveying
National Council of Structural Engineers Associations
National Society of Professional Engineers
Society of Allied Weight Engineering, Inc.
Society of Fire Protection Engineers
Society of Manufacturing Engineers
Society of Naval Architects and Marine Engineers
Society of Women Engineers
Standards Engineering Society
Tau Beta Pi Association, Inc.
The American Ceramic Society
The Minerals, Metals & Materials Society
The Society of Petroleum Engineers
United Engineering Foundation

INDUSTRY

Abbott Laboratories
Adobe Systems
Advanced Micro Devices, Inc.
The Aerospace Corporation
American Electric Power Company, Inc.
Amgen Inc.
Anadarko Petroleum Corporation

Baxter International Inc.
Becton, Dickinson and Company
Bethlehem Steel Corporation
The Black & Decker Corporation
The Boeing Company
Caterpillar Inc.
Conectiv
Conoco Inc.
Corning Incorporated
The Dow Chemical Company
DTE Energy Company
Duke Energy Corporation
E. I. du Pont de Nemours and Company
Eastman Chemical Company
Eastman Kodak Company
Edison International (Southern California Edison Co.)
Emerson
Exxon Mobil Corporation
FedEx Corporation
Fluor Corporation
FMC Corporation
Ford Motor Company
Foster Wheeler Ltd.
General Electric Company
General Motors Corporation
Georgia-Pacific Corporation
The Goodyear Tire & Rubber Company
Honeywell International Inc.
Illinois Tool Works Inc.
Intel Corporation
International Business Machines Corporation
ITT Industries, Inc.
Jacobs Engineering Group Inc.
Johnson & Johnson
KeySpan Corporation
Kinder Morgan, Inc.
Lennox International Inc.
Lockheed Martin Corporation
Minnesota Mining and Manufacturing Company
Motorola, Inc.
Niagara Mohawk Holdings
Northeast Utilities
OGE Energy Corp.

Owens Corning
Parker Hannifin Corporation
Phelps Dodge Corporation
Phillips Petroleum Company
Progress Energy, Inc.
Seagate Technology, Inc.
Thermo Electron Corporation
Thornton-Tomasetti Group
TRW Inc.
United Technologies Corporation
USX Corporation
Valero Energy Corporation
The Walt Disney Company
The Williams Companies, Inc.
Xerox Corporation

DESIGN AND CONTRACTING FIRMS

Bechtel Group, Inc.
Black & Veatch
Camp Dresser & McKee Inc.
CH2M Hill Companies Ltd.
The Clark Construction Group
Duke Engineering & Services
HNTB Companies
The Louis Berger Group Inc.
Parsons Brinckerhoff Inc.
Parsons Corp.
Sargeant & Lundy LLC

COLLEGES AND UNIVERSITIES

University of Alabama
Arizona State University
Boston University
University of California – Berkeley
University of California – Davis
University of California – Irvine
University of California – Santa Barbara
California Institute of Technology
Carnegie Mellon University
University of Cincinnati
University of Colorado at Boulder

APPENDIX F

Florida Institute of Technology
University of Florida – Gainesville
Georgia Institute of Technology
Harvard University
Harvey Mudd College
University of Illinois at Urbana-Champaign
Illinois Institute of Technology
Indiana University/Purdue University at Indianapolis
University of Iowa
Johns Hopkins University
Kansas State University
Lehigh University
Louisiana State University
University of Louisville
University of Maine
University of Maryland
University of Massachusetts – Amherst
University of Massachusetts – Lowell
Massachusetts Institute of Technology
Miami University
University of Michigan – Ann Arbor
University of Michigan – Dearborn
University of Missouri – Columbia
New Mexico State University
University of New Mexico
State University of New York – Buffalo
North Carolina State University
Ohio State University
University of Oklahoma
Pennsylvania State University
University of Pennsylvania
University of Pittsburgh
Purdue University
University of Rhode Island
Rochester Institute of Technology
Rose-Hulman Institute of Technology
Smith College
University of Southern California
Stanford University
Texas A&M University
Tufts University
University of Utah
Vanderbilt University

University of Vermont
Villanova University
Virginia Polytechnic and State University
University of Virginia
University of Washington
Western Michigan University
University of Wisconsin – Madison
University of Wisconsin – Milwaukee

INDUSTRY ASSOCIATIONS

American Forest & Paper Association
American Petroleum Institute
Cellular Telecommunication & Internet Association (CTIA)
National Association of Manufacturers
National Mining Association
Telecommunications Industry Association

EDUCATION ASSOCIATIONS AND FOUNDATIONS

Alfred P. Sloan Foundation
American Council on Education
Association for Educational Communications and Technology
Association of American Colleges and Universities
Council for Advancement and Support of Education
Council of Chief State School Officers
National Association of Colleges and Employers
National Association of State Universities and Land-Grant Colleges
Women in Engineering Programs and Advocates Network (WEPAN)

NATIONAL LABORATORIES

Argonne National Laboratory
Bettis Laboratory
Brookhaven National Laboratory
DOE Joint Genome Institute
Fermi National Accelerator Laboratory
Idaho National Engineering and Environmental Laboratory
Lawrence Livermore National Laboratory
Los Alamos National Laboratory
National Energy Technology Laboratory
National Renewable Energy Laboratory
Oak Ridge National Laboratory

Pacific Northwest National Laboratory
Sandia National Laboratories
Savannah River Technology Center; Westinghouse Savannah - River Company

FEDERAL CONTACTS

Defense Advanced Research Projects Agency
Federal Highway Administration
NASA (Chief Engineer; Office of Public Affairs)
National Science Foundation
U.S. Department of Defense
U.S. Environmental Protection Agency

PRODUCERS/MEDIA

Thirteen/WNET
Production Group, Inc.
Palfreman Film Group
RDF Media

MUSEUMS

Arizona Science Center, Phoenix, AZ
Carnegie Science Center, Pittsburgh, PA
Great Lakes Science Center, Cleveland, OH
Institute for Exploration, Mystic, CT
Liberty Science Center, Liberty State Park, Jersey City, NJ
Miami Museum of Science & Space Transit Planetarium, Miami, FL
MOSI - Museum of Science & Industry, Tampa, FL
Museum of Science, Science Park, Boston, MA
National Air and Space Museum, Smithsonian Institution
North Carolina Museum of Life and Science, Durham, NC
Reuben H. Fleet Science Center, San Diego, CA
SciWorks, The Science Center & Environmental Park of Forsyth County,
    Winston-Salem, NC
The Exploratorium, San Francisco, CA
The Tech Museum of Innovation, San Jose, CA